INNOVATIONS
IN HOME
ENERGY USE

A Sourcebook for Behavior Change

INNOVATIONS IN HOME ENERGY USE

A Sourcebook for Behavior Change

BRIAN G. SOUTHWELL, ELIZABETH M. B. DORAN, AND LAURA S. RICHMAN

EDITORS

RTI Press

The RTI Press mission is to disseminate information about RTI research, analytic tools, and technical expertise to a national and international audience. RTI Press publications are peer-reviewed by at least two independent substantive experts and one or more Press editors.

RTI International is an independent, nonprofit research organization dedicated to improving the human condition by turning knowledge into practice. RTI offers innovative research and technical services to governments and businesses worldwide in the areas of health and pharmaceuticals, education and training, surveys and statistics, advanced technology, international development, economic and social policy, energy and the environment, and laboratory testing and chemistry services.

Library of Congress Control Number: 2015958583

ISBN 978-1-934831-15-1
(refers to print version)

RTI Press publication No. BK-0015-1512
http://dx.doi.org/10.3768/rtipress.2015.bk.0015.1512
www.rti.org/rtipress

Cover design: Dipali Aphale

This publication is part of the RTI Press Book series.

RTI International

3040 East Cornwallis Road, PO Box 12194, Research Triangle Park, NC 27709-2194, USA

rtipress@rti.org

www.rti.org

About the Editors

Brian G. Southwell directs the Science in the Public Sphere program in the Center for Communication Science at RTI International and holds faculty appointments at Duke University (through Duke's Energy Initiative) and at the University of North Carolina at Chapel Hill. At Duke, he was faculty adviser for Project LIT HoMES (Leveraging Individual Transitions into Homeownership to Motivate Energy Savings), which inspired this book. He also hosts the public radio show *The Measure of Everyday Life* on WNCU.

Elizabeth M. B. Doran is an environmental engineer and has been a doctoral candidate in Earth and Ocean Sciences at Duke University throughout the time this book was written. At Duke, she also led the steering committee for the Project LIT HoMES Summit that led to this book.

Laura S. Richman is a faculty member in Psychology and Neuroscience at Duke University. As a faculty participant in the Bass Connections project at Duke, she co-led Project LIT HoMES with Dr. Southwell.

Contents

Preface

You can trace the lineage of this book directly to the Project LIT HoMES (Leveraging Individual Transitions into Homeownership to Motivate Energy Savings) Summit, which was hosted by a team of Duke University and RTI International collaborators in February 2015 on the campus of Duke University. After months of planning and initial conversations, a wide range of professionals converged at Duke to discuss and refine innovations intended to curb homeowner energy use in the United States. Following that summit, we invited participants and additional collaborators to develop a series of essays that explain various innovative ideas.

As a collection of essays that explore innovations to encourage reduction in homeowner energy use, this volume reflects a confluence of ideas and initiatives from a variety of perspectives rather than a narrow look at what a single, particular line of academic literature suggests might be possible to shape homeowner behavior. Not only do the contributors represent a wide array of institutions and backgrounds, but the very intellectual infrastructure that encouraged and allowed the summit that inspired this book itself represents a conscious effort to facilitate multidisciplinary and interdisciplinary collaboration for the purpose of addressing salient societal concerns.

The effort—a summit that brought together people from research institutions, county sustainability offices, government agencies, consultant organizations, architecture firms, building contractors, and real estate agencies—required a physical home and tools for planning and collaboration. Duke University's Bass Connections program, and specifically the Bass Connections in Energy program led by Dr. Richard Newell, offered that home and those resources for both the day-long summit and the broader year-long Project LIT HoMES effort. What follows in these pages is a gathering of thought and development of ideas that aligns well with many of the goals of the Bass Connections program, which founders intended to be a university-wide effort to link faculty and students in order to address complex challenges through problem-focused education.

We hope that this book offers a foundation for new dialog about ways in which homeowners can be engaged as partners in the quest to reduce our collective energy use. Undoubtedly, individual behavior is only one part of our energy consumption equation, but it nonetheless is a noteworthy one and a potentially changeable element. People generally do not want to waste energy wantonly, and many people likely find the idea of saving energy to be reasonably attractive. The question lies in how to change actual behavior, which appears to be a function of not only general attitudes but also perceptions of social norms, specific skill knowledge, and available technology and tools.

This book suggests that the fruitful pathways toward a future of reduced energy use at home wind through sectors of professionals and practitioners who often do not converse. For this reason, we have fashioned a conversation here that can be inspirational as an interdisciplinary model. We will need such innovation if we are to live responsibly on the planet in coming decades.

Brian Southwell
April 2015
Durham, North Carolina

Acknowledgments

Because of the foundational perspective of the project as a teaching and learning exercise, the core team who organized this summit was a faculty-student collaborative operating at Duke University in 2014 and 2015 comprising team leaders Brian Southwell and Laura Richman and team members Elizabeth Doran, Brandon Ellis, Nicholas Garafola, Lauren Harper, Amit Singhpal, and Jordan Thomas. That team, organized under the title Project LIT HoMES (or Leveraging Individual Transitions into Homeownership to Motivate Energy Savings), first spent time surveying available literature on human behavior and energy efficiency and savings interventions and then turned to the question of how best to organize people to generate and develop new ideas. We moved away from the idea of hosting a conventional conference with a lineup of presentations and instead developed the idea of hosting an idea incubator summit modeled on similar efforts at Boston's Museum of Science and at RTI International in Research Triangle Park, North Carolina.

Early consultants who were helpful in developing our ideas included Mishel Filisha of the New York State Energy Research and Development Authority, Erika Shugart of the American Society for Microbiology, David Sittenfeld of the Museum of Science, Boston, and Troy Livingstone and Brad Herring of Durham, North Carolina's Museum of Life and Science, as well as Mayme Webb-Bledsoe, Kyle Bradbury, and Steve Hicks of Duke University. Suky Warner of the Duke University Energy Initiative also provided useful logistical help. As these ideas came into fruition and we developed a book project, a number of staff at RTI International and RTI Press offered valuable insight, including Karen Lauterbach, Dorota Temple, Jacqueline Olich, Joanne Studders and Sonja Douglas. We are grateful for all of the support from RTI International and RTI Press.

The summit could not have occurred without the earnest contributions of all participants. In addition to people already named above, those participating in person at the summit or in other ways online as we planned for the event included Charles Adair, Steve Armstrong, Ellis Baehr, Gibea Besana,

Kyle Bradbury, Phillip Bradrick, Katie Bray, Joshua Burton, Megan Carroll, Donna Coleman, Laurie Colwander, Daniel Conner, Drew Cummings, Dan Curry, Erin DeBerardinis, Satana DeBerry, Jason Elliott, Tobin Freid, Jeff Furman, Christopher Galik, Melanie Girard, Karan Gupta, Nathan Hsieh, Tim Johnson, Daniel Kauffman, Nikhil Kaza, Jim Kirby, Laura Langham, Barbara Levine, Maria Mauceri, Ryan Miller, Andrea Ortiz, Sarah Parvanta, Jon Poehlman, Casey Quinn, Doug Rupert, Danielle Sass Byrnett, Larry Shirley, Claudia Squire, Evans Taylor, Joseph Threadcraft, Debby Warren, Jennifer Weiss, Michael Youth, and Marina Ziemian.

Introduction

This volume touches on several salient points of intervention where a better understanding of human behavior can impact the greater landscape of energy conservation in the residential context. Growing out of a particular space and time, these chapters are intended to provide stimulus for future action by practitioners and students alike. These ideas are grounded in the practical experience of the authors and data from supporting case studies. Collectively they address the clear need for improved understanding of the human behavior interface in enacting change.

Part I of this book focuses on providing context. Chapter 1 discusses the multiple levers for change in the context of home energy use and lays out a case for the importance of the human behavior lever. Chapter 2 turns to framing the human behaviors specifically. To do so, chapter author Jordan Thomas revisits and reframes a taxonomy of the core behaviors relating to energy use in the home, laying the groundwork for further discussion of interventions.

The book then transitions to Part II and the role of energy knowledge on behavior. This section begins with a detailed discussion in Chapter 3 of the economic implications and rationale for investing in energy efficiency and conservation in the residential sector. The discussion by chapter authors Daniel Kauffmann and Nicholas Garafola illustrates the challenges that exist in calculating value in the residential context. The focus moves to a detailed analysis of the potential role of energy information in shifting behavior with a focus on the home buying process in Chapter 4. Many people begin with Internet-based real estate searches to find a home to buy or rent, so efforts to assess and improve the online availability of information about various homes' energy use could be an important step in improving subsequent resident energy use. Chapter 5 addresses information sharing and dissemination, particularly in the neighborhood context. Based on evidence built from experience running neighborhood education programs, the authors offer concrete advice and demonstrate real value in an often underappreciated strategy for change.

The book concludes with a trio of chapters focused on opportunities for residential energy engagement in Part III. Chapter 6 looks at the opportunities for leveraging synergies in two established levers for homeowner engagement: retrofits and risk-reduction activities. Chapter 7 explores the synergistic opportunity for employers and employees to leverage mutual benefit from home energy reductions. Specifically, based on experience building a program at a large educational institution, the authors discuss the barriers faced by employees and how employers can ameliorate such barriers while receiving mutual benefit in the form of offsets as an example. Finally, Chapter 8 concludes this section and the book with a look at how to keep homeowners engaged in the process of making change happen. Based on experience from large-scale research efforts, this chapter offers concrete strategies for practitioners and policymakers alike.

The diversity of chapters in this book is intentional. With this arrangement, we argue for an interdisciplinary approach that considers and includes the role of human behavior among other factors that affect energy use. The chapters that follow as part of this volume contribute novel ideas to a nascent intersection of scholarship and professional practice that offers a useful path forward.

PART I
INTRODUCTORY CONSIDERATIONS: ENERGY BEHAVIOR IN CONTEXT

CHAPTER 1

Leverage Points for Achieving Sustainable Consumption in Homeowner Energy Use

Elizabeth M. B. Doran

Fundamental to the human condition is the persistent quest to satisfy certain basic requirements, including water, food, and shelter. This has been true for the whole of human history, and for the whole of human history these endeavors have required the expenditure of energy. Whereas that energy could at one time have been described accurately as "manpower," and later "horsepower," energy consumers are largely removed today from the process of energy generation. Particularly in the developed world, we enjoy a level of convenience in this regard that allows the vast majority of us to spend our time in pursuit of other endeavors rather than searching for power for basic life functions. This convenience, however, comes with tradeoffs, including increased environmental impact and generally entrenched demand that requires persistent attention to the supply of raw materials. It is in the context of this convenience that we should consider efforts to manage and shape home energy use.

Below, I outline the scope of the challenge and the available avenues to address it. The levers of technology, policy, and macro-level economic incentives are popular with many officials and professionals. Efforts to decrease demand, increase efficiency, and reduce impact are diverse and are expanding to include efforts to affect home resident energy use behavior through creative means such as communication interventions and organizational partnerships.

The Challenge

The collective impact of the energy required to create and sustain the modern domestic experience has large-scale implications. On the production side of the equation, a great deal of effort goes into the process of securing the persistent and growing supply of the energy that reaches our homes. At the same time, waste byproducts are having an outsized impact on the natural functioning of the Earth system. For the past 40 years, the US Energy

Information Agency (EIA) has been compiling and providing statistics on the production and consumption of energy in the United States.

The agency reports that, since 1950, residential energy consumption has increased from 17 percent of the total energy consumed in the country to 22 percent (EIA, 2015a). In addition to becoming a larger proportion of the energy used, the total energy consumed has also increased, with total residential consumption having grown 260 percent (EIA, 2015a). In the last decade, consumption has grown a more modest 2.4 percent, but the ratio of domestic to imported energy has shifted dramatically. In 2004, domestic production was enough to satisfy 70 percent of domestic demand; by 2014, that number had risen to 88 percent (EIA, 2015b). The result of an "all of the above" shift in domestic energy policy to enable more domestic production (US Department of Energy, 2014), this has been accomplished through efficiency; new methods of production, including hydraulic fracturing for natural gas production, that have opened harder-to-produce resources; and expanded implementation of renewable energy technology. Pursuit of a secure energy supply will remain a national priority and security interest for the foreseeable future (Brown et al., 2014).

The environmental impact of energy consumption is also a concern. In 2009, the residential sector accounted for approximately 20 percent of the total greenhouse gas emissions of the United States (EIA, 2015c,d). As the US is the second largest national emitter of greenhouse gasses in the world, this contribution alone is not insignificant. Greenhouse gas accumulation in the atmosphere is known to contribute to the observed average warming of the Earth's atmosphere and the changing dynamics of the ocean-atmosphere system that creates the Earth's climate (Intergovernmental Panel on Climate Change [IPCC], 2013). These changes pose potentially dire consequences to the habitability of the planet in the long term, are a concern for national security in the medium term, and threaten domestic infrastructure in the short term.

These reasons alone could be sufficient to want to reduce overall energy consumption and increase the efficiency of home energy use in particular. However, an additional and perhaps more germane rationale for seeking reductions, particularly in the residential space, is the financial cost. In this regard, there are gains to be made on both the individual and national levels. The EIA estimates that since the turn of the century, Americans spend on average roughly 3 percent of household income on energy, or nearly $2,000 per year (EIA, 2015e). Although there is wide variability in this number, that equates to a roughly $230 billion residential energy sector in 2015.

Reducing homeowner energy use reduces expense. In economically burdened households, this alone can have significant implications for well-being (Bertrand et al., 2006). On a larger scale, reduced demand ameliorates the need to build more capacity, although by how much depends on the cost of new energy generation, which varies widely. To use the most extreme example, if the capital required to build a nuclear power plant were used instead to install solar arrays, the result would be 147 times the energy-generating capacity (EIA, 2013).

Meanwhile, as the day on which you are reading this began, millions of American citizens went about their morning routines: flipping a switch to turn on a light, powering up an electric toothbrush, putting on a fresh change of clothes, checking a cell phone, starting a coffee maker, preparing toast in a climate-controlled home. Each seemingly simple and mindless action was powered by electricity. Although most of these actions more than meet the requirements for survival, they also are nevertheless intimately intertwined with the provision of relatively basic needs. It is this complex interaction of infrastructure, expectation, and behavior that makes up the true context for ongoing efforts to reduce the use of energy in the home.

Traditional Avenues to Energy Reduction

As easy as energy currently is to access in most of the United States and other countries in the (economically) developed world, the task of reducing energy use is surprisingly complex. This is true despite the several obvious avenues one could pursue to reduce use such as technology improvements, economic incentives, and policy interventions. These three avenues indeed have all received much attention and achieved significant success at reducing home energy use and warrant discussion here to set the stage for behavioral intervention as a strategy.

Technology Improvements

Improving technology arguably offers the most seductive solution to energy reduction. It is, after all, the point of contact for energy use in the first place. Take the lightbulb: it is the exemplar of this phenomenon, as its sole purpose is to turn energy into light. Heat, in this case, is energy wasted. A better lightbulb, then, will use less energy to provide the same amount of light and not get as hot as the original incandescent bulb. This can be done and has been done in numerous ways with more efficient lightbulb technology currently flooding the market. Those new technologies, however, did not just appear without direct and planned effort. LED lightbulbs, for instance, are the result of significant research and development that took decades of technical innovation to achieve. The result is a product that lasts significantly longer and uses a fraction of the energy, as little as one-tenth, as the previous-generation product. Similar advances have been made for other household technologies, including washing machines, refrigerators, and water heaters.

If household appliances are typically replaced after about a decade, the lifespan of the house itself is an order of magnitude longer. Since space conditioning (heating and cooling) is also typically the largest consumer of home energy use across the country, the importance of building design, material choice, and construction methods becomes that much more important. This includes decisions about a home's size, the amount and type of insulation used in walls and floors, and the orientation of the home on the land and in relation to the sun. These are all choices made at the beginning of the home-building process. Improvements in material technology and building methods are thus another large area of focus for research and development efforts.

Achieving greater efficiency is just one metric by which to measure the energy use of a product or home. While energy efficiency is the most directly important to the total sum of a monthly energy bill and the area we focus on here, the capital energy, or the energy used to create and transport a product, has been increasingly under scrutiny as well. Life cycle analysis is a method for determining the impact of a product over the course of its sourcing, manufacture, distribution, use, and disposal—its life cycle. Attention to the life cycle of products and materials reminds us that turning a technology on and off for its primary use is not the only way in which a product designed for the home uses energy. It can also tell us just how important the efficiency of the product is relative to its overall impact and this turns out to be quite significant (Boustani et al., 2010). Of course, the assumptions of use used in such a modeling effort can have a significant impact on the findings and are an important reason to consider human behavior.

Economic Incentives

The second lever is economics, a measure of the societally agreed-upon value of goods and services in society. The most basic tool in the economics toolbox is return on investment (ROI) calculations. Used in the sale of a new, more efficient technology, for example, the capital cost is balanced against the savings realized at the new operating cost relative to the cost of operating the existing technology. ROI is typically reported as a unit of time, i.e., years or months. These numbers are often used to both justify newer technology options and sell consumers on them. But the method gets hung up, particularly in situations where the payback period is longer than the planned use of the technology.

Let us return to the case of the LED lightbulb discussed above. The advertised lifespan of a new bulb might be 10 or 20 years. If a consumer purchases the lightbulb and then moves, the consumer misses a huge chunk of the energy savings promised in that ROI calculation. In this case, the consumer must enter into another economic calculation: the cost-benefit analysis. We could contrive a scenario in which, say, the consumer doesn't have a lightbulb, and therefore the benefit of having light is greater than not making any investment at all, and the LED lightbulb is the only option available at the box store—a reality that may be true in the future. The point is that more sophisticated economic levers also exist to encourage behavior change. In addition to cost-benefit analysis that might take into consideration more factors than just the ROI, economics offers numerous other incentive structures including variable pricing and discounts or tax breaks to encourage desired market behavior.

Variable pricing is designed to shift energy consumption that might otherwise happen at peak hours of the day, due to convenience, to off-peak hours, when base load power is more readily available and cheaper to generate. This works particularly well in the energy sector because of the way power is generated. Electricity producers typically operate continuously at what is called a base load level. This base load power might be a coal-fired or nuclear power plant, generators that typically take a long time or expense to start up and shut down and are therefore most efficient if operated continuously. An optimization on the part of a utility determines how much base power to generate, and then that energy is available regardless of demand fluctuations, sometimes in excess.

During periods of high demand, however, say in the evening and on hot summer afternoons, demand will outstrip this supply. Utilities accomplish the demand match by supplementing the base load with a more expensive, but also more responsive, production technology. When utilities can't meet demand, they have to employ alternative strategies like rolling blackouts. However, when there is excess base load power capacity during periods of low demand, utilities want to sell that energy, and they can sell it more cheaply both because it is cheaper to generate and because it is otherwise wasted. Variable pricing encourages a smearing of demand by charging less for electricity when it is cheaper to generate it and more when it is more expensive. The setup is clearly desirable for the producer of electricity, and the consumer has the opportunity to pay less as well. Applications where this might work particularly well for consumers include plugging in an electric car to charge overnight.

Discounts are another example of an economic policy instrument designed to spur demand. Tax breaks are an example of a discount and have been applied to electric vehicles, home purchasing, solar technology installations, and the purchase of some Energy Star–certified home appliances and technologies. In the context of a technology purchasing decision, like the lightbulb example from earlier, a discount will effectively lower the capital cost of the technology and the length of time it takes to realize the ROI through energy savings. In theory this should make the decision to purchase the technology more attractive and therefore more likely.

While economic assessments and levers are highly successful, one of the fundamental shortcomings of them is the basic assumption of a rational consumer. A rational consumer is one who maximizes the utility of purchases given known budget constraints. The assumption sets up nice graphs and formulas and allows for economic analysis that is rational and calculable. However, in making a solvable problem, the rational consumer assumption ignores important realties about the motivating forces driving human decision making. This is a strong and oft-repeated criticism of economic levers. Understanding the limitations and opportunities at this interface is clearly an opportunity to improve deployment and adoption measures.

Policy Interventions

Finally, policy interventions are often a compliment to the levers of technology and economics. Let us return once again to the lightbulb example. In 2007, the US federal government adopted an efficiency standard for lightbulbs

(US Energy Independence and Security Act, 2007). The policy indicated that it would be illegal to sell products that did not meet the adopted standard. The result was the effective phaseout of the incandescent lightbulb from the marketplace. While inefficient, the incandescent lightbulb was incredibly cheap to manufacture and sell, and for the most part lightbulb producers were unable to sell more efficient bulbs that were more expensive. As a result of the phaseout, consumers no longer had the option to purchase the inefficient technology; this created an instant market for the newer, more efficient technology.

For lightbulb manufacturers, this is an example of a command and control policy instrument. Other applications of this type of rule-making in the energy sector include fuel standards for vehicle fleets and the elimination of lead from gasoline. Command and control policy relies heavily on a clear understanding of the impact of the undesirable behavior so an appropriate limit and penalty can be applied. This is true in part because reaching and ensuring compliance can then be a significant burden.

More common is an alternative approach to command and control that uses the power of a relatively free market to either determine the optimal technological solution or distribute the burden of a desired level of regulation. This can include setting standards, but not specifically mandating how the standard is met. In the lightbulb example, efficiency of the product category, not the type of technology, was regulated despite the effective result of eliminating the incandescent bulb from general purpose applications. Also in that case, a phased approach was utilized so the market for high wattage lightbulbs was affected first, then lower and lower wattage lightbulbs until all incandescent bulbs were effectively eliminated. This allowed manufacturers to phase out old products while phasing in new ones.

A hybrid approach of this nature was used in the control of acid rain–causing emissions from power plants across the country; in that case, a tradable permit system was added. Called a cap and trade program, regulators issued "permits to pollute" and allowed utilities to trade among themselves. Utilities that could easily cut their emissions were able to sell their unused permits in a marketplace. Over time, the number of permits has been reduced.

Such programs have shifted the impact of generation activities while other programs have made similar impacts on consumption. Building energy codes, for example, are established at the state or local zoning level and are applied to new construction and renovations. Municipalities that lack the bandwidth to develop their own unique codes typically adopt standards from national research organizations like ASHRAE, the American Society of Heating, Refrigerating and Air-Conditioning Engineers. This is an example of the interface of policy with technology. Established as a strictly voluntary program, Energy Star is another. The Energy Star program is operated by mandate under the Clean Air Act by the US Environmental Protection Agency and was originally established in 1992. The federally mandated program seeks to promote energy efficiency through the measurement and reporting of energy consumption and cost data. It effectively operates as an oversight of the claims made by manufacturers in the appliance and built environment marketplaces. Products must be third-party verified by approved laboratories, and are then spot-checked off the shelf to verify claims. The intent of this validation is clearly directed at changing behavior both in manufacturing and purchasing.

This discussion of the levers of technology, economics, and policy should clearly indicate the impact each can have on the consumption of electricity. And yet it would also be fair to recognize that they each come up against human factors of decision making and behavior that are outside their scope of influence.

A Role for Human Behavior

Residential energy use continues to grow despite attention from academia, governments, and industry to each approach reviewed above. This growth can only be partly explained by the increase in the total number of households. According to data available from the US Census, the total number of households in the US has grown 270 percent between 1950 and 2010, while data available from the EIA indicates the total consumption of energy in the residential sector has grown 365 percent over the same time period (EIA, 2015a; US Census Bureau, 2003, 2012). One must conclude that, over time, American households have grown to use increasing levels of energy despite efforts to the contrary, and new strategies must be deployed to continue to make progress against this growing demand.

One such avenue that has received relatively little attention in the context of home energy use is the role of human behavior in shaping demand. In fact, a recent review of the energy literature by Sovacool (2014), including more than 9,500 authors and 90,000 references from three leading energy journals over 15 years, documents the nascent level of the contributions of the social sciences to the curated body of literature. Areas most heavily addressed include the technology of energy production—generation, transmission, and distribution—with nearly a quarter of the articles addressing some aspect of that process. The most frequently addressed topic areas included energy markets, public policy mechanisms, climate change, and pricing. In total, those four topics accounted for more than half of the articles reviewed. In contrast, the areas of land use, behavior, and research and development were found least frequently, accounting for less than 5 percent of the total literature reviewed. While the findings were likely influenced by the journals chosen, they nevertheless suggest a gap. This could be viewed as problematic or fortunate since it likely represents fertile ground for further research and means there is a wide opportunity to explore new avenues and strategies for reduction.

The called-for shift in the social science agenda joins a growing movement set in motion by researchers addressing sustainability more broadly and sustainable consumption specifically. Both movements stem from international efforts dating as far back as the 1980s to articulate the connection between the environment and human development, as well as an ambition for the future of this dual relationship (United Nations World Commission on Environment and Development, 1987). Inclusion of the human dimension in global environmental change research, however, has taken decades to actualize. Fomenting in the latter part of the 20th century (National Research Council, 1999) and explicitly called for at the dawn of the millennium (Kates et al., 2001), sustainability science is the dominant organizing disciplinary movement, with associated arms of research and pedagogical learning. Sustainability science is still in its normative phase, but it fundamentally seeks to simultaneously address issues of intra- and intergenerational equity, resource provision, and Earth system function. Germane to the topics addressed here, these concerns collide, for example, at the intersection of the energy-water-food nexus (Bazilian et al., 2011), where the security of these life-sustaining provisions is paramount. Even this one framing leaves the science's purview flexibly broad, if also somewhat ambiguous.

Embedded within the growing calls for a science of sustainability, in the early decade of the 2000s, the social science fields turned their attention to addressing the emerging concern that consumption was to blame for the modern environmental crisis. Early efforts to address the emerging agenda of what was dubbed "sustainable consumption" held forth a similar critique to the one presented here, namely, that a strictly techno-economic approach to policy and action would limit the ability of society to address what was, in fact, a question of expectation, normality, well-being, and institutionalization of demand (Southerton et al., 2004). In making this distinction at the time, Southerton and colleagues gave appropriate separation between the human dimensions of consumption and a parallel technical approach dubbed dematerialization that prioritized material considerations of efficiency and substitution over functional considerations (van der Voet et al., 2003)—what we have here dubbed the lever of technology intervention.

Not specific to energy, sustainable consumption research efforts nevertheless laid several foundational elements that reemerge in the context of our present concern with residential energy consumption. For example, in outlining an argument against neoclassical economic assumptions of rational consumer choice, Southerton and colleagues (2004) describe the collective, normative, and routine constraints on consumers that limit rational choice. These can include what Shove (2003), writing on the same subject in her now highly cited work, describes as "conventions of comfort, cleanliness, and convenience." Both pieces focus on the practice of showering, but a common and important conclusion is the argument that a holistic, systems approach is necessary to understand the particular metrics of consumption. This remains true in the energy space as well.

Researchers focused on sustainable consumption have also begun to grapple with the evaluation of programs designed to encourage sustainable or green lifestyles through information and educational campaigns. One such effort in the United Kingdom called Action at Home was evaluated by Hobson (2001) and appears to offer an early evaluation of the usefulness of information stemming from recognition that barriers to action may not be limited to physical or infrastructural problems but also to the decoupling of awareness and action. Again, this remains true in the energy space (Southwell et al., 2012).

Sustainable consumption continues to be a salient field of research with a maturing methodological approach that also faces systemic barriers to further progress including the availability of integrative data for deeper insights (Dietz, 2014). These barriers apply equally to the behavior and energy use agenda, but are beginning to be addressed. Early estimates indicate a 20 percent reduction potential for residential energy use through behavioral intervention (Dietz et al., 2009). While a review of almost 40 published home energy use reduction intervention studies indicated the promise of several intervention strategies despite methodological issues (Abrahamse et al., 2005).

Efforts to shape the social science agenda in the energy space offer numerous pathways forward. In establishing a new journal dedicated to the pursuit of integrating social science into the energy arena, Sovacool (2014) offers an extensive sample of possible topics and research questions that might be pursued. With 14 topics and 75 questions offered, the list is extensive. The topical areas include the need for increased analysis using human-centered research methods, and the need for interdisciplinary approaches despite the known difficulty in conducting such research. Because energy can be a gender-centered activity, particularly in poor and developing parts of the world, the role of gender and identity is included as well, with questions such as "What constitutes 'gender-aware' energy planning?" Sovacool calls on philosophy and ethics to address the political and moral questions related to the fair distribution and use of energy in a world of limits that further considers the distribution of the costs and benefits, as well as the implications for future generations. Several additional topics, which are germane to the chapters that follow here, include communication, social psychology, behavior, anthropology and culture.

The Way Forward

In beginning to address the wide ambitions put forth in the agendas for sustainability, sustainable consumption broadly, and energy use in particular, the editors of this volume took particular note of the observation by Sovacool (2014) that there is a disconnect between the published authorship and energy managers and practitioners in the field. Moving beyond the efforts to bring the social sciences to bear on behavior change, this volume adopts a collaborative, interdisciplinary, and practical approach that began with a summit of professionals and has resulted in this volume of ideas and strategies for practice. This is an ambition of the science of sustainability as well and while interdisciplinary efforts will likely remain difficult to implement, they are nonetheless necessary components of the practice of science in the service of society (Lubchenco, 1998).

What about the prospects for influencing change? Emeritus Professor Gene Rochlin's (2014) retrospective look at the role of the social sciences in the climate change debate may offer some insight. Written initially in 1989 and then published 25 years later, the piece offers an accounting of the perception of the social sciences in the broader debate about climate change and our energy future. In particular, Rochlin notes that the contributions of the social sciences, specifically political science, often seem general, reactive, and cautious in their findings and recommendations, which may seem particularly frustrating to policymakers and activists, as well as ineffective. This is due to numerous factors including disciplinary norms and mismatch in the expectations and modes of operation of the numerous parties.

A companion review of the past 25 years, however, also suggests that despite these limitations, real contributions have emerged from energy research conducted within the social sciences (Ryan et al., 2014). Of note here is the fact that social science has successfully informed a broader understanding of the biases humans hold in risk evaluation and decision making by individuals and the impact this has on policy outcomes. In seeking to apply this knowledge and activate what Ryan et al. call energy citizenship, the following chapters might just represent micropractices that lead to microdemographic shifts; small efforts distributed widely can nonetheless add up to large change over time.

Chapter References

Abrahamse, W., Steg, L., Vlek, C., & Rothengatter, T. (2005). A review of intervention studies aimed at household energy conservation. *Journal of Environmental Psychology, 25*, 273–291.

Bazillian, M., Rogner, H., Howells, M., Hermann, S., Arent, D., Gielen, D., ... Yumkella, K. K. (2011). Considering the energy, water and food nexus: Towards an integrated approach. *Energy Policy, 39*(12), 7896–7906.

Bertrand, M., Mullainathan, S., & Shafir, E. (2006). Behavioral economics and marketing in aid of decision making among the poor. *Journal of Public Policy and Marketing, 25*(1), 8–23.

Boustani, A., Sahni, S., Graves, S. C., & Gutowski, T. G. (2010). Appliance remanufacturing and life cycle energy and economic savings. In *Proceedings of the 2010 IEEE International Symposium on Sustainable Systems & Technology*, IEEE, 2010. 1–6.

Brown, M. A., Wang, Y., Sovacool, B. K., & D'Agostino, A. L. (2014). Forty years of energy security trends: A comparative assessment of 22 industrialized countries. *Energy Research & Social Science, 4*, 64–77.

Dietz, T. (2014). Understanding environmentally significant consumption. *PNAS: Proceedings of the National Academy of Sciences of the United States of America, 111*(4), 5067–5068.

Dietz, T., Gardner, G. T., Gilligan, J., Stern, P. C., & Vandenbergh, M. P. (2009). Household actions can provide a behavioral wedge to rapidly reduce US carbon emissions. *PNAS: Proceedings of the National Academy of Sciences of the United States of America, 106*(44), 18452–18456.

Hobson, K. (2001). Sustainable lifestyles: Rethinking barriers and behavior change. In M. J. Cohen & J. Murphy (Eds). *Exploring sustainable consumption: Environmetnal policy and the social sciences* (pp. 213–224). Oxford, UK. Elsevier Science Ltd.

Intergovernmental Panel on Climate Change (IPCC). (2013). *Climate change 2013: The physical science basis.* Contribution of Working Group I to the fifth assessment report of the Intergovernmental Panel on Climate Change [T. F. Stocker, D. Qin, G.-K. Plattner, M. Tignor, S. K. Allen, J. Boschung, . . . P. M. Midgley (Eds.)]. Cambridge, UK and New York, NY: Cambridge University Press.

Kates, R. W., Clark, W. C., Corell, R., Hall, J. M., Jaeger, C. C., Lowe, I., . . . Svedin, U. (2001). Environment and development. *Sustainability science. Science, 292*(5517), 641–642.

Lubchenco, J. (1998). Entering the century of the environment: A new social contract for science. *Science, 279*(5350), 491–497.

National Research Council. (1999). *Our common future: A transition toward sustainability.* Washington, DC: National Academies Press.

Rochlin, G. I. (2014). Energy research and the contributions of the social sciences: A retrospective examination. *Energy Research & Social Science, 3,* 178–185.

Ryan, S. E., Hebdon, C., & Dafoe, J. (2014). Energy research and the contributions of the social sciences: A contemporary examination. *Energy Research & Social Science, 3,* 186–197.

Shove, E. (2003). Converging conventions of comfort, cleanliness and convenience. *Journal of Consumer Policy, 26,* 395–418.

Southerton, D., Van Vliet, B., & Chappells, H. (2004). Introduction: Consumption, infrastructures and environmental sustainability. In D. Southerton, H. Chappells & B. Van Vliet (Eds.), *Sustainable consumption: The implications of changing infrastructures of provision* (pp. 1–11). Cheltenham, UK: Edward Elgar Publishing.

Southerton, D., Warde, A., & Hand, M. (2004). The limited autonomy of the consumer: Implications for sustainable consumption. In D. Southerton, H. Chappells & B. Van Vliet (Eds.), *Sustainable consumption: The implications of changing infrastructures of provision* (pp. 32–48). Cheltenham, UK: Edward Elgar Publishing.

Southwell, B. G., Murphy, J. J., DeWaters, J. E., & LeBaron, P. A. (2012). *Americans' perceived and actual understanding of energy.* RTI Press Publication No. RR-0018-1208. Research Triangle Park, NC: RTI Press. http://dx.doi.org/10.3768/rtipress.2013.bk.0011.1306

Sovacool, B. K. (2014). What are we doing here? Analyzing fifteen years of energy scholarship and proposing a social science research agenda. *Energy Research & Social Science, 1,* 1–29.

United Nations World Commission on Environment and Development. (1987). *Report of the World Commission on Environment and Development: Our common future* (Annex to document A/42/427). Retrieved from http://www.un-documents.net/wced-ocf.htm

US Census Bureau. (2003). Households by type and size: 1900 to 2002 [Table]. In *US Census Bureau, Statistical Abstract of the United States: 2003.* Retrieved from https://www.census.gov/statab/hist/HS-12.pdf

US Census Bureau. (2012). Households, families, subfamilies, and married couples: 1980 to 2010 [Table]. In *US Census Bureau, Statistical Abstract of the United States: 2012.* Retrieved from http://www.census.gov/compendia/statab/cats/population/households_families_group_quarters.html

US Department of Energy. (2014). *US Department of Energy strategic plan 2014–2018* (DOE/CF-0067). Retrieved from http://energy.gov/downloads/2014-2018-strategic-plan

US Energy Independence and Security Act. (2007). Public Law No. 110-140, 121 Stat. 1492.

US Energy Information Agency. (2013). Updated estimates of power plant capital and operating costs [Table 1]. In *Capital costs for electricity plants: Updating capital cost estimates for utility scale electricity generating plants.* Retrieved from http://www.eia.gov/forecasts/capitalcost/

US Energy Information Agency. (2015a). Energy consumption by sector [Table 2.1]. In *Monthly Energy Review April 2015.* Retrieved from http://www.eia.gov/totalenergy/data/annual/index.cfm#consumption

US Energy Information Agency. (2015b). Primary energy overview [Table 1.1]. In *Monthly Energy Review April 2015.* Retrieved from http://www.eia.gov/totalenergy/data/annual/index.cfm#consumption

US Energy Information Agency. (2015c). Carbon dioxide emissions from energy consumption by source [Table 12.1]. In *Monthly Energy Review April 2015.* Retrieved from http://www.eia.gov/totalenergy/data/annual/index.cfm#consumption

US Energy Information Agency. (2015d). Carbon dioxide emissions from energy consumption: Residential sector [Table 12.2]. In *Monthly Energy Review April 2015.* Retrieved from http://www.eia.gov/totalenergy/data/annual/index.cfm#consumption

US Energy Information Agency. (2015e). Summary household site consumption and expenditures in the US—totals and intensities, 2009 [Table CE1.1]. In *2009 RECS survey data*. Retrieved from http://www.eia.gov/consumption/residential/data/2009/index.cfm?view=consumption#summary

Van der Voet, E., van Oers, L., & Nickolic, I. (2003). *Dematerialization: Not just a matter of weight* (CML report 160). Netherlands, Center for Environmental Science, Leiden University. Retrieved from http://www.cml.leiden.edu/research/industrialecology/research/publications-ie-00-04s.html

—

CHAPTER 2

EVALUATING THE THEORETICAL JUSTIFICATION FOR TAILORED ENERGY INTERVENTIONS:
A Practice-Oriented Analysis of an Energy-Relevant Behavior Taxonomy

Jordan Thomas

Introduction

As industrialized countries strive to achieve their emissions reduction goals, many public entities and utilities are encouraging homeowners to adopt energy-saving behaviors. The average homeowner is currently able to make cost-effective decisions in the home that would significantly reduce energy consumption with little to no change in his or her lifestyle. Even so, household energy-saving behavior is lagging behind the available technology, creating a gap between the current consumption level and a cost-effective level of consumption (known as the energy efficiency gap; Jaffe & Stavins, 1994). Thus arises the question, What is keeping energy consumers from reducing their energy consumption? If cost-effectiveness is not appealing enough, how can public entities best motivate homeowners to adopt energy-saving behaviors?

Given the complicated nature of human behavior, it is no surprise that there is little agreement on how to best motivate energy-saving behavior. Utilities and government bodies have implemented a wide variety of behavioral interventions, yielding a variety of results. While some principles of intervention design seem to be unilaterally effective, such as presenting the intervention from a reliable source and making the process of participation simple (Dietz et al., 2009; Dietz et al., 2013), policymakers disagree on which psychological determinants of behavior deserve the most focus.

One option is to target human attitudes and preferences. Abrahamse and colleagues (2005) refer to this strategy as "voluntary behavior change." Within this option lies numerous other options: does the policymaker go about encouraging change by providing information or by making normative statements about energy consumption? If the policymaker choses to disseminate information, how will it be framed—in terms of cost-effectiveness or environmental harm? In contrast with voluntary behavior change, another option is to change the context in which energy decisions are made. Abrahamse refers to this strategy as making "structural changes," that is, to provide incentives, rebates, or rewards for the behaviors one wants to encourage. Of course, these options are not mutually exclusive and many interventions often employ multiple strategies at once.

The type of psychological determinants that a program designer choses to employ depends on the designer's definition of "energy-saving behavior" and understanding of the human decision making process. There are two primary schools of thought regarding the definition of energy-saving behavior. The first is that all energy-saving behaviors result from the same set of psychological determinants, making energy-saving behavior a unidimensional, holistic behavior. This theory is closely related to the theory that pro-environmental behaviors are an aggregate behavior and should be treated as such (Kaiser, 1998). Accordingly, proponents for this school of thought favor comprehensive, one-size-fits-all interventions. The second school of thought assumes that energy-saving behavior consist of many different types of behavior, each differing in its psychological determinants (Stern, 2000; Black et al. 1985). If this is true for human behavior, then an intervention that appeals to a single determinant of behavior (such as cost savings, for example) would encourage only the behaviors subject to that particular determinant and leave the others unchanged. Accordingly, this school of thought favors interventions that are tailored to the type of behavior that is being encouraged. Recently, many studies on household energy behavior have made findings that support this second model and call for more nuanced interventions (Botetzagias et al., 2014; Dietz et al., 2009; Kowsari & Zerriffi, 2011; Urban & Ščasný, 2012).

Herein lies the debate: Is it sufficient to address household energy behavior as one single construct or variable, thus simplifying the dilemma of complex intervention design, or is it reasonable to believe that maximum energy reduction can only be achieved by using sets of more nuanced, targeted, or tailored interventions that acknowledge multiple types of behavior? If it is the latter, then how targeted or tailored do the interventions need to be to effectively encourage energy savings? Must there be an intervention for each individual energy-saving behavior, or can the behaviors be targeted as groups?

The purpose of this chapter is to evaluate the theoretical justification for tailored interventions by comparing the psychological determinants of energy-saving behaviors. The assumption behind this evaluation is that if a behavior, or group of behaviors, depends on a similar set of psychological determinants, then those behaviors can be activated with a single intervention, making intervention tailoring unnecessary. I rely on literature that Karlin and colleagues used in their 2014 review of household energy behavior dimensions as well as more recent studies identified to be relevant to the research question. Although Karlin and colleagues used this literature to look for a consensus on how behaviors fall into groups, I use the literature to look for a consensus on how groups of behaviors are motivated.

Background

A Taxonomy of Household Energy-Saving Behaviors

In order to understand how nuanced interventions differ from unidimensional ones, one must first understand to what the interventions are being tailored—i.e., the different types of household energy-saving behaviors. Karlin et al. (2014) made the most recent and most thorough effort made to establish a taxonomy of behaviors. They created the taxonomy by coding 28 studies and papers that either reference or explicitly study how energy-saving behaviors relate to one another. Their review found that the majority of the studies that differentiate between dimensions of behavior agree on two main clusters of behaviors: curtailment and investment. Table 2-1 offers a summary of Karlin et al.'s results.

Table 2-1. Two clusters of household energy behaviors

	Curtailment behavior	Investment behavior
Cost	Free/Low	Low/High upfront investment
Frequency	High	Low
Actions	Usage	Structural
Permanence	Low	High
Lifestyle	Loss of comfort	No change or improvement
Impact	Low	High
Population	All	More difficult for renters
Motivation	Moral	Financial/Moral
Examples	Turning off appliances when they are not in use. Lowering the temperature of the water heater. Only using the washing machine for a full load.	Insulating the attic or weatherizing the home in general. Purchasing energy-efficient appliances.

Tailoring Interventions

I assessed behavioral interventions that seek to alter the adoption and use of technology so that residential energy consumption decreases. An effective intervention should be persuasive to the target audience or at least supportive of the intended behavior, meaning intervention strategy should focus on appropriate motivators of the behavior being encouraged. Examples of psychological determinants that are thought to affect energy-saving behavior include perceived loss of comfort from energy saving, concern for energy as an issue, price concern, personal responsibility to save energy, and positive environmental attitude.

Methods

I used a variety of methods to identify relevant studies. Primarily, because Karlin and colleagues' research question is closely analogous to my research question, the studies used in their 2014 review, "Dimensions of Conservation: Exploring Differences Among Energy Behaviors," served as the primary source of literature for my review. In addition, I identified relevant papers using searches of JSTOR, PsycINFO, and Web of Science databases. In order for a study to be included in the analysis, it had to make a conclusion about which psychological determinant plays a role in determining a specific energy-saving behavior (or type of behavior). For each paper, I recorded the behavior(s) considered and the psychological determinant affecting that behavior. Then, for each type of behavior, I compared the sets of identified determinants and, if the determinants were distinct, judged that type of behavior to be sufficiently distinct from other types, thereby justifying a tailored intervention.

Results

I identified 28 relevant studies, 26 of which were used in the Karlin et al. (2014) literature review. Studies differed greatly in their relevance to the research question. While all of them mentioned at least some form of taxonomy of energy-saving behaviors, many measured changes in household energy consumption over time as opposed to changes in behavior (Abrahamse et al., 2005; Ayers et al., 2009; Curtis et al., 1984; Stern & Gardner, 1981; Cialdini & Schultz, 2004). By doing so, the researchers failed to measure how the various types of behaviors differ in their reaction to interventions. In a similar vein, some studies measured changes in types of behaviors but did not measure the determinants of those behaviors (Dietz et al., 2009; Kempton et al., 1992;

Lehman & Geller, 2004; Stern, 1992). This reaffirms Abrahmase and colleagues' (2005) finding that underlying psychological determinants of energy-saving behaviors are not well understood. In addition, some papers that did consider the determinants of types of behaviors did so only through conjecture instead of analyzing new empirical evidence (Oikonomou et al., 2009; McKenzie-Mohr, 1994).

The studies that did focus on the determinants of energy-saving behaviors can be grouped into two categories. The first category ("Category A") comprises papers that studied the determinants of energy-saving behaviors as a whole (Barr et al., 2005; Curtis et al., 1984; Oikonomou et al., 2009). These studies found that placing importance on social obedience over social power (Barr et al.), having a personal norm for energy conservation (Curtis et al.; van der Werff & Steg, 2015), and having a sense of efficacy (Oikonomou et al.) are all psychological determinants of energy-saving behavior. The second category ("Category B") comprises papers that studied the determinants of specific types of energy-saving behaviors. The majority of these papers used the two-group taxonomy explained in the Background section of this chapter: curtailment or investment. The determinants for curtailment behaviors were found to be personal norm for energy curtailment (Black et al., 1985), descriptive norm for energy conservation (Cialdini & Schultz, 2004; Macey & Brown, 1983), and high environmental concern (Urban & Ščasný, 2012). The determinants for investment behaviors were placing importance on reducing energy use, belief that one's energy cost is high (Nair et al., 2010), and high environmental concern (Urban & Ščasný).

One paper, by Black and colleagues (1985), used narrower groups of behaviors instead of the two-group taxonomy of curtailment or investment. Black and colleagues found there to be different determinants for high-cost investment and low-cost investment. High-cost investment was found to be predicted by perceived self-interest, while low-cost investment was predicted by a personal norm for energy efficiency.

Discussion

The literature review yielded a complex array of results. All energy-saving behaviors may share at least a set of psychological predictors. Namely, Urban and Ščasný (2012) found that environmental concern is a predictor for both curtailment and investment behaviors. Multiple papers made claims about determinants of energy-saving behavior as a whole (the Category A papers), but they are not useful for this review because they did not test the determinants of behavior types separately, meaning which behaviors are being affected by the observed determinants is unknown.

Although the psychological determinants of behavior types overlap somewhat, there also seem to be distinct sets of determinants for curtailment and investment behaviors. Curtailment behaviors seem to be more sensitive to personal and social norms, while investment behaviors may be more sensitive to cost considerations.

As made evident by the Black et al. (1985) study, it is possible that the two-group taxonomy misses important differences between high- and low-cost investment behaviors.

In terms of designing behavior interventions, it is unclear whether the differences between groups of behaviors are distinct enough to warrant being treated by separate, tailored interventions. It seems possible that a single intervention that activates personal norms for environmental concern (Van der Werff & Steg, 2015) may encourage the adoption of all types of energy-saving behaviors. Conversely, it may be prudent for an intervention to be tailored specifically to investment behaviors because consumers tend to underestimate the savings of investment behaviors and investment behaviors tend to yield higher energy savings than curtailment behaviors (Attari et al., 2010; Gardner & Stern, 2008).

Given the inconclusive results of the literature review, policymakers can consider other factors affecting the cost-effectiveness of tailored interventions. First, tailoring interventions to determinants of narrow groups of behaviors adds to the programmatic cost of the intervention. Thus, if it is unclear whether tailoring will yield significantly improved energy savings, then that cost should be avoided. Second, the practicality of this study is limited to demographics that are not constrained by contextual determinants of behavior. The phrase "contextual determinant" refers to factors that are outside of the consumer's control (Barr et al., 2005; Nair et al., 2010).

Black and colleagues (1985) found that the strongest contextual determinant of high-cost investment behaviors was homeownership, and the strongest contextual determinant of curtailment behaviors was direct payment for home heating. Thus, the psychological determinants that an intervention appeals to do not matter for renters who are unable to act on those psychological motivations. These contextual obstacles may be circumvented by using interventions that target owners of the multifamily buildings.

This study attempted to reveal the theory behind the best practices of intervention design. However, the human decision-making process is complicated, making it difficult to generalize any consensus in the literature. Another approach to understanding effective intervention design would be to conduct randomized trials on homeowner behavior. Such trials have been completed before, but most do not take into consideration the psychological determinants being activated by the intervention nor the type of behavior being targeted. Making these considerations explicit in the documentation of randomized trials would yield a stronger understanding of how behavior reacts to various intervention strategies.

Chapter References

Abrahamse, W., Steg, L., Vlek, C., & Rothengatter, T. (2005). A review of intervention studies aimed at household energy conservation. *Journal of Environmental Psychology, 25,* 273–291.

Attari, S. Z., DeKay, M. L., Davidson, C. I., & Bruine de Bruin, W. (2010). Public perceptions of energy consumption and savings. *Proceedings of the National Academy of Sciences of the United States of America, 107,* 16054–16059.

Ayres, I., Raseman, S., & Shih, A. (2009). *Evidence from two large field experiments that peer comparison feedback can reduce residential energy usage* (NBER Working Paper Series, Vol. w15386). Retrieved from http://ssrn.com/abstract=1478804

Barr, S., Gilg, A. W., & Ford, N. (2005). The household energy gap: Examining the divide between habitual- and purchase-related conservation behaviors. *Energy Policy, 33,* 1425–1444.

Black, J. S., Stern, P. C., & Elworth, J. T. (1985). Personal and contextual influences on household energy adaptations. *Journal of Applied Psychology, 70,* 3–21.

Botetzagias, I., Malesios, C., & Poulou, D. (2014). Electricity curtailment behaviors in Greek households: Different behaviors, different predictors. *Energy Policy, 69,* 415–424.

Cialdini, R. B., & Schultz, W. (2004). *Understanding and motivating energy conservation via social norms* (Tech. Rep.). Menlo Park, CA: William and Flora Hewlett Foundation.

Curtis, F., Simpson-Housley, P., & Drever, S. (1984). Household energy conservation. *Energy Policy, 12,* 452–456.

Dietz, T., Gardner, G. T., Gilligan, J., Stern, P. C., & Vandenbergh, M. P. (2009). Household actions can provide a behavioral wedge to rapidly reduce US carbon emissions. *PNAS: Proceedings of the National Academy of Sciences of the United States of America, 106,* 18452–18456.

Dietz, T., Stern, P. C., & Weber, E. U. (2013). Reducing carbon-based energy consumption through changes in household behavior. *Daedalus, 142*(1), 78–89.

Gardner, G. T., & Stern, P. C. (2008). The short list: The most effective actions US households can take to curb climate change. *Environment: Science and Policy for Sustainable Development, 50*(5), 12–25.

Jaffe, A. B., & Stavins, R. N. (1994). The energy-efficiency gap: What does it mean? *Energy Policy, 22*(10), 804–810.

Kaiser, F. G. (1998). A general measure of ecological behavior. *Journal of Applied Social Psychology, 28*, 395–422.

Karlin, B., Davis, N., Sanguinetti, A., Gamble, K., Kirkby, D., & Stokols, D. (2014). Dimensions of conservation: Exploring differences among energy behaviors. *Environment and Behavior, 46*(4), 423–452.

Kempton, W., Darley, J. M., & Stern, P. C. (1992). Psychological research for the new energy problems. *American Psychologist, 47*, 1213–1223.

Kowsari, R., & Zerriffi, H. (2011). Three dimensional energy profile: A conceptual framework for assessing household energy use. *Energy Policy, 39*(12), 7505–7517.

Lehman, P. K., & Geller, E. S. (2004). Behavior analysis and environmental protection: Accomplishments and potential for more. *Behavior and Social Issues, 13*, 13–32.

Macey, S., & Brown, M. (1983). Residential energy conservation: The role of past experience in repetitive household behavior. *Environment & Behavior, 15*, 123–141.

McKenzie-Mohr, D. (1994). Social marketing for sustainability: The case of residential energy conservation. *Futures, 26*, 224–233.

Nair, G., Gustavsson, L., & Mahapatra, K. (2010). Factors influencing energy efficiency investments in existing Swedish residential buildings. *Energy Policy, 38*, 2956–2963.

Oikonomou, V., Becchis, F., Steg, L., & Russolillo, D. (2009). Energy saving and energy efficiency concepts for policy making. *Energy Policy, 37*, 4787–4796.

Stern, P. C. (1992). What psychology knows about energy conservation. *American Psychologist, 47*, 1224–1232.

Stern, P. C. (2000). Toward a coherent theory of environmentally significant behavior. *Journal of Social Issues, 56*, 407–424.

Stern, P. C., & Gardner, G. T. (1981). Psychological research and energy policy. *American Psychologist, 36*(4), 329.

Urban, J., & Ščasný, M. (2012). Exploring domestic energy-saving: The role of environmental concern and background variables. *Energy Policy, 47*, 69–80.

Van der Werff, E., & Steg, L. (2015). One model to predict them all: Predicting energy behaviours with the norm activation model. *Energy Research & Social Science, 6*, 8–14.

PART II
ENERGY KNOWLEDGE AND BEHAVIOR

CHAPTER 3

Quantifying the Value of Home Energy Improvements

Daniel Kauffman and Nicholas Garafola

Introduction to Measuring Energy Savings from Home Improvements

If you have recently performed an energy-saving improvement on your home, such as replacing windows, sealing air leaks, or installing new insulation, you probably spent a great deal of money in the hope that you will save even more money over time through lower energy bills. This expenditure is analogous to buying a bond, where money is spent up front to buy the bond in order to receive periodic interest payments from owning the bond over time. It is also similar to, and in fact the opposite of, paying down mortgage or credit card debt. Though there are benefits to investing in home energy improvements beyond energy savings over time, such as an increase in home value and improvements in comfort and indoor health, if the primary motivator for the investment is energy savings, then money spent on improvements should be viewed as an alternative use of capital to debt repayment or traditional investment assets.

An economist or financial advisor would advise a homeowner that, in hindsight, a home energy improvement was a good investment if the rate of return on that investment as measured by the value of energy saved over time is greater than the homeowner's borrowing rate, for example the mortgage rate. Otherwise, the money would have been better spent paying off the mortgage.

If indeed you recently spent money on an energy-saving improvement, you are probably well aware of how much money you spent. You may have also noticed that your utility bills have gone down, or maybe you have not. Was the home energy improvement in fact a good investment? In order to quantify the value of the savings and therefore the improvement, a homeowner must first measure the amount of energy saved.

Energy Evaluation, Measurement, and Verification

The process of measuring energy savings involves comparing how much energy has been consumed to how much energy would have been consumed had the energy-saving improvement not taken place. Quantifying counterfactuals is inherently imprecise, and approximating what would have happen had the retrofit not taken place presents a fundamental problem that cannot be resolved through simple measurement.

Energy evaluation, measurement, and verification (EM&V or M&V) is the process used by energy engineers, utility consultants, and others to determine the quantity of energy saved as a result of an energy conservation measure. EM&V is a broad field applicable to both utility-driven energy efficiency programs and consumer-side energy improvements, and to a range of facility types including commercial buildings, industrial facilities, and single-family dwellings.

Broadly speaking, EM&V practitioners use select among four general methods to determine energy savings, referred to here by their naming conventions in the Efficiency Valuation Organization's International Performance Measurement & Verification Protocol (International Performance Measurement & Verification Protocol Committee, 2002):

- Option A: Retrofit Isolation—Key Parameter Measurement
- Option B: Retrofit Isolation—All Parameter Measurement
- Option C: Whole Facility Measurement
- Option D: Calibrated Simulation

Option A: Retrofit Isolation—Key Parameter Measurement

The process of retrofit isolation involves measuring or estimating energy savings from the replacement of a specific piece of equipment with a more efficient one. Commercial buildings and industrial facilities have many pieces of replaceable equipment, such as pumps, fans, motors, chillers, boilers, and lighting, many of which are independently monitored. The energy savings due to replacing such equipment can be determined by measuring or estimating the operating hours and power draw of the equipment prior to replacement, and then replicating measurement or estimation for the new equipment after replacement. This option is generally considered acceptable when the effects of replacing a single piece of (or in the case of lighting, set of) equipment can be accurately estimated or measured without significant uncertainty in the operating characteristics, and the continued effectiveness of the equipment can be reassessed through simple retesting over time.

Retrofit isolation with key parameter measurement is generally not suitable for quantifying the value of a given home's energy improvements because few home energy improvements involve a discrete piece of equipment (i.e., an isolatable retrofit) for which the energy consumption is independently measured. One exception could be a refrigerator that is always plugged in. A simple plug load power draw measurement prior to replacement can be compared to the plug load power draw measurement of a replacement refrigerator, and the measurement can be repeated periodically to detect degradation in performance over time. Absent a power measurement, the nameplate power ratings of the old and new refrigerators can be compared (note that quantifying energy savings by comparing power ratings before and after equipment replacement and extrapolating operating hours would, technically speaking, be considered Retrofit Isolation—No Parameter Measurement). However, the example of the refrigerator is unique in that a refrigerator is always plugged in, and so the operating hours can be easily extrapolated. For other appliances, the operating hours would either have to be measured and logged though appliance monitoring, communicating plug load monitors, or branch circuit monitoring in the circuit breaker or be computed though an algorithmic technique such as signal decomposition of high-resolution whole home energy consumption data.

The effect of replacing or repairing a home's HVAC (heating, ventilation, and air conditioning) system can also, in theory, be isolated, but doing so is highly problematic. Very rarely is the electrical consumption of an HVAC system independently measured over time, nor are its operating hours logged. The output of an air conditioning system can be determined by measuring its energy efficiency ratio, representing the ratio of cooled, dehumidified output air of the air conditioning system to input electrical energy required to cool and dehumidify the air at any given time (measured in British thermal units per kilowatt-hour, Btu/kWh). However, measuring energy savings by doing so is unusual for multiple reasons. First, few HVAC systems have their energy efficiency ratio measured upon commissioning, and in even fewer cases is the energy efficiency ratio of the old HVAC system being replaced measured for baseline reference. A comparison of nameplate seasonal energy efficiency ratios (SEER), though indicative of temperature- and humidity-dependent real-time operating energy efficiency ratio, is not sufficient for accurate determination of energy savings from HVAC system replacement and at best can project an estimate. Many HVAC systems do not operate at their optimal energy efficiency ratios due to errors in commissioning, such as insufficient coolant charge or unsynchronized fan motor and compressor motor speeds. Furthermore, degradation in HVAC system performance due to age, wear, lack of maintenance, tampering, or other effects is difficult to account for in performance estimates. Unless the input and output of an HVAC system is monitored over time, its performance cannot be reliably estimated for the purposes of accurately quantifying ongoing energy savings.

Finally, the effects of improvements to the home's envelope, such as adding or replacing insulation, sealing air leaks, recaulking or replacing windows, and other improvements typically provided by home contractors, cannot be isolated because such improvements are not individual appliances whose energy consumption can be measured (note that a building's envelope is the physical separation between the indoor climate-conditioned space within the building and the ambient environment). For measuring the improvements to a home's envelope, only options C and D below can be reasonably considered.

Option B: Retrofit Isolation—All Parameter Measurement

Option B is similar in methodology to option A but requires the measurement of all parameters pertaining to the energy consumption of replaced appliances. Measurement of all parameters is preferable in situations where energy usage monitors are being deployed anyhow. One example would be where metering is included with the equipment, such as the case where an outside party is responsible for all of the energy saved. Though such a situation may arise in a commercial building, it would nearly never arise in a house, and the difficulty and cost associated with the measurement of all relevant parameters for a home improvement would be prohibitively difficult and costly and altogether unnecessary. Whereas measurement of key parameters for home appliance replacement is generally unsuitable, measurement of all parameters for home appliance replacement is generally impractical.

Advances in measurement technology and the placement of energy usage sensors in home appliances may make accurate retrofit isolation possible in the future, but so long as the power draw and operating time of home appliances are not regularly recorded, accurate retrofit isolation will remain an impractical way to quantify energy savings for anything but discrete applianc replacements.

Option C: Whole-Facility Measurement

Whole facility (or whole-building) M&V involves an analysis of the utility usage data of the building. Also known as the main-meter approach, energy savings is quantified by comparing the pre- and post-retrofit utility meter readings (often from utility bills) to determine the difference between what was consumed post-retrofit and a projection of what would have been consumed had the retrofits not occurred. Determining the projection of hypothetical usage involves the calculation of a whole-building energy use model of the home before the retrofit, where the whole-building energy use model is a statistical model relating the energy use of the home over a given period of time and one or more independent variables, such as outdoor temperature. As reduced utility bills are the paramount indicator of energy savings, whole building energy use modeling with weather correction is in fact the only means available to a homeowner to truly quantify the realized value of home improvements that involve envelop upgrades such as insulation, air sealing, windows, and the like.

Whole-building energy use modeling is not without challenges. The whole-facility approach is most appropriate when expected savings is significant compared to whole-home energy use, otherwise the energy savings may be difficult to discern from the random variations in month-to-month usage within a home. Also, implicit in a whole-building energy use model is that, absent the improvement, the home is in substantially similar operation in the pre- and post-retrofit measurement periods. Changes in occupancy (assuming that occupancy is not considered as an independent variable) or the addition or subtraction of a major energy-consuming appliance outside of the retrofit invalidates the implicit assumption of operational similarity that is core to the whole-facility approach. Though potentially imprecise, whole-building energy use modeling is the best approach available to a homeowner to quantify energy savings from home improvements.

The technical standard for whole-building energy use modeling is ASHRAE-14, "Measurement of Energy and Demand Savings" (ASHRAE, 2002). ASHRAE-14 details a whole-building approach in which monthly consumption of electricity and natural gas is evaluated as a function of average outdoor temperature during the interval between meter reads.

A commonly used alternative to average outdoor temperature is heating and cooling degree days, where degree days are a measure of how hot or cold a given period of time is relative to a base temperature. For example, given a base heating temperature of 65 degrees Fahrenheit, if the average temperature in a 30-day winter billing period were 55 degrees, that billing period would have 300 heating degree days (HDD) associated with it (i.e., (65 - 55 degrees) × 30 days). Though 65 degrees has commonly been used as a base temperature for both heating and cooling, using 60 degrees for heating and 70 degrees for cooling usually provides a better statistical fit. Ideally, the base temperatures will be variable rather than fixed in the model, as variable-base models reveal additional information about the home's energy consumption for heating, cooling, and non-temperature-related uses. Where a complete set of metered (rather than estimated) utility bills is available, ASHRAE-14 recommends the use of at least 12 months of data in the pre-retrofit period, at least nine post-retrofit data points, and the computation of a regression model that meets a minimum criterion.

Though several best-fit model statistics are available, the recommended statistical test is the determination of the coefficient of variation of the model's root mean squared error, or CV(RMSE). A CV(RMSE) of less than 20 percent is generally sufficient for a model to be considered valid, and a lower CV(RMSE) and therefore a better-fit model can be achieved by, for example, varying the degree day bases.

In April 2013, the National Renewable Energy Laboratory (NREL) released the *Uniform Methods Project: Methods for Determining Energy Efficiency Savings for Specific Measures* (Jayaweera & Haeri, 2013) to serve as a set of generally accepted standard practices within the M&V profession. Included in the Uniform Methods Project report (as chapter 8) is an update to the whole-facility method: "Whole-Building Retrofit with Consumption Data Analysis Evaluation Protocol" (Agnew & Goldberg, 2013). For measuring energy savings on a given home, Agnew and Goldberg recommend use of the following five-parameter model for electricity consumption (pp. 8-13–8-14):

$$E_m = \mu + \beta_H H_m + \beta_C C_m + \varepsilon_m$$

where:

E_m = average consumption per day during interval m

μ = average daily baseload consumption estimated by the regression

H_m = specifically, $H_m(\tau_H)$, average daily heating degree days at the base temperature(τ_H) during meter read interval m

C_m = specifically, $C_m(\tau_C)$, average daily cooling degree days at the base temperature(τ_C) during meter read interval m

β_H, β_C = heating and cooling coefficients estimated by the regression

ε_m = regression residual

The Uniform Methods Project recommends calculating a coefficient of electric heating for all homes, including natural gas–heated homes, as some heating-related electricity use is expected to be consumed regardless of heat source, for example, to power the air circulation fan in the HVAC system. For homes heated with natural gas, there is no need to compute a cooling coefficient of natural gas use, and so a three-parameter model of natural gas use is sufficient.

Option D: Calibrated Simulation

A calibrated simulation is a computer simulation of the energy use of replacement equipment within a home or of the whole home. In such a simulation, parameters of the home and the improvements made are entered, pre-retrofit usage is modeled and calibrated to actual historic usage, and then post-retrofit savings are forecasted based on the simulation. Air infiltration values collected through blower door tests and insulation R-values are common parameters used to estimate a home's energy use profile through simulation.

Calibrated simulations can be used to quantify the value of home improvements in situations where no pre-retrofit data are available, for example when evaluating different insulation, fenestration, and other envelope options while constructing a new home (though strictly speaking, without actual utility data, the model itself is not calibrated). The technique can also be used where no post-retrofit data are available or significant baseline energy adjustments have been made, for example with an expansion or demolition that changes the area inside the home. In such cases, a comparison of pre- and post-retrofit utility bills cannot usefully measure the impact of improvements to a home because, from a structural perspective, the pre- and post-retrofit utility bills reflect two different homes. Also, the technique can be useful where additional improvements are to be made during the evaluation period following initial improvements such that sufficient data cannot be collected during the interim to determine savings from the initial improvement.

Though a useful tool in estimating future savings, a calibrated simulation is not a preferred means of quantifying energy savings from improvements made to an existing home. A calibrated simulation absent post-retrofit data (utility bills, equipment inspection, etc.) is in fact a form of prediction rather than measurement. Also, a properly calibrated simulation requires considerable skill to accurately perform. Such simulations are rarely performed after the fact and are simply not necessary for quantifying savings if utility bills can readily be analyzed using a whole-facility model. However, the combined use of both a whole-facility model and a calibrated simulation can provide insight into not only the value of the home improvements based on their energy savings, but also why exactly the home saved energy and which measures had what effect. Whereas retrofit isolation is a bottom-up approach involving adding the impact of individual improvements to estimate energy savings, the combination of whole building modeling and calibrated simulation is a top-down approach that measures savings and allocates those savings to individual improvements.

According to the *Energy Management Handbook* (Doty & Turner, 2012), the reference manual published by the Association of Energy Engineers, a general procedure for selecting an M&V approach (for any facility) can be summarized as follows:

1. Try to perform monthly utility bill before/after analysis.

2. If this does not work, then perform daily or hourly before/after analysis.

3. If this does not work, then perform component isolation analysis.

4. If this does not work, then perform calibrated simulation analysis.

5. Report savings and finish analysis.

For a homeowner, what matters in improving home energy efficiency is paying lower utility bills for substantially the same energy services. In the context of quantifying savings from home energy improvements, the Association of Energy Engineers–recommended procedure simplifies to what all homeowners already know by intuition: to quantify energy savings, look closely at the utility bills.

Why Measure Energy Savings, and Why Not

Despite the available methods, measurement of energy savings from home improvements is in fact rarely performed, and therefore the value of home energy improvements is rarely known. There are several reasons why energy savings measurement is so rare. As explained, energy saved is not a number that is simply recorded over time into a database as utility meter data are and the power produced by solar panels are (though in the authors' opinion, if a homeowner has undertaken an energy savings retrofit that was paid for in part through a tax credit, a utility rebate, or other public assistance, as almost all retrofits are, such a number should in fact be entered over time into a database for the sake of public accountability). But the main reason why energy saved is not typically measured today is because nobody has a strong vested interest in measuring it.

What about home contractors? Contractors are certainly interested in knowing how much energy they save their client homeowners, but not enough so to go out of their way to measure it. Contractors are paid up front regardless of whether or not energy is in fact saved, and some contractors have concerns that if the work that they do for clients is overly focused on energy savings, which can be quantified, the scope of work may exclude other worthwhile improvements whose value can be even harder to quantify, such as improving the healthiness of the living environment, the structural integrity, and the overall quality of the home. Contractors are professionals who are generally very good at following industry best practices and guidelines when performing their work, especially work that can be visually audited. But energy savings is not typically audited, and so properly measuring energy savings is not a best practice that many contractors are trained to follow.

Until contractors are compensated based in part upon the amount of energy saved over time, contractors will not have a strong interest in measuring energy savings. In fact, contractors have a disincentive to quantify energy savings: doing so calls into question the accuracy of the energy savings estimates presented to the homeowner prior to commencement of the project. The actual impact of this effect is not well known (presumably because energy savings are so rarely quantified), but according to one study of 13 contractors in California, an average of only about 34 percent of projected gas savings from jobs completed by contractors was actually realized by homeowners following the work (Energy Upgrade California, 2015). Undoubtedly, many contractors would like to measure energy savings from their work to provide potential clients with both more precise estimates and also credible references that the savings that they estimate are in fact realized. However, contractors are not so interested that they would spend their valuable time collecting such information. Notwithstanding, if quantifying energy savings were easy and cheap (or mandatory) for contractors to perform, it stands to reason that many, especially the most reputable ones, would in fact do so.

What about the government? Local, state, and federal government agencies often subsidize home energy improvement projects, most often through the tax code or through government grant programs. The most widespread government-subsidized home energy efficiency programs are weatherization assistance programs, whose purpose is to assist homeowners in lowering their utility bills and making their homes more livable. Eligibility for weatherization assistance is dependent on both economic factors, such as income, and also home-specific factors, such as energy use intensity.

As with other non-homeowner stakeholders, the government does not directly accrue benefit from the improvements in proportion to the energy savings, and so precise measurement is not required. Providing low-income families with improved homes while also creating jobs for contractors and local tradespersons are sufficient reasons to implement weatherization assistance programs. Return on investment is a secondary concern, as the investment is the public's and the return is the recipient homeowner's. Public home energy efficiency programs sometimes collect and analyze data from a sample of homes in order to better understand program impacts, ensure quality work, comply with reporting requirements, and other purposes, but such collection and analysis is typically performed by weatherization programs on only a small fraction of participating homes, if at all.

What about utilities? Electric and natural gas utilities run home energy efficiency programs themselves, such as giving away LED lightbulbs and subsidizing HVAC repair. Though it may seem counterintuitive that a utility that profits from a home using more of its product would help a homeowner to not use so much of it, utilities have reasons for wanting homeowners to consume less energy.

Electric utilities invest money in the capacity to produce and distribute the electricity needed to supply their customers, and in many instances those customers pay the same flat rate for every kilowatt-hour they buy. But each kilowatt-hour does not cost a utility the same to provide. During the middle of the night when demand is low, plenty of electricity is available, and wholesale electricity prices (i.e., the price of electricity capacity that electricity distributors pay) tend to be low. In the middle of the day in mid-summer when everyone wants to operate their air conditioners at the same time, electricity is scarcer and many utilities lose money on each electron, as they provide expensive electricity to their customers while charging them the same rate they always do. Such utilities would generally prefer that homeowners buy less of these money-losing peak demand kilowatt-hours and more of the cheaper off-peak variety (though this effect is less pronounced in places with time-of-use pricing, as part of the additional cost is passed on to the customers).

Utilities that compete for homeowners' business as retail distributors of electricity can offer energy efficiency programs to their customers to improve customer loyalty in the hope that by helping to keep homeowners' bills low the homeowners will be apt to renew their contracts with that electricity distributor instead of switching to a competitor; such programs are also offered as promotional tools for new customer recruitment. Natural gas utilities have an interest in homeowners' using natural gas to efficiently heat their homes in winter because the alternative would be electric heating, and so many natural gas utilities offer energy efficiency programs to help enhance and promote the efficient use of natural gas. As with all industries, competition helps to keep prices down for customers.

In many cases, regulated utilities receive cost recovery and even profit from their energy efficiency programs (by raising electricity rates for everyone) if they can satisfactorily show regulators that their energy efficiency programs are indeed saving energy. This cost recovery is achieved by submitting an EM&V report to the state utility commission for approval. These EM&V reports are usually authored and audited by independent consultants so as to avoid conflicts of interest within the utility that would arise if it appeared that the utility was reporting savings while simultaneously selling electricity that would not have been consumed in the first place.

Though these EM&V reports seek to estimate energy savings, they do so on a program-wide basis and not on a home-by-home basis. The savings estimated within these M&V reports are stipulated on the basis of work to be performed for the amount of energy that ought to be saved. For example, installation of LEDs in place of incandescent bulbs ought to save a certain amount of energy over time, and utilities can receive compensation on the basis of simply providing LEDs to homeowners. Such savings are known as *deemed savings* rather than *measured savings*.

Though a sample of homes might have energy savings measured by utilities as a check on the stipulations and to adjust and restate predicted savings after the fact, by no means are all or even most homes subject to savings measurement over time. Another approach used by utilities, known as *pooled savings*, involves estimating a large group of homes' energy savings by comparing their gross usage over the course of a utility energy efficiency program period to the energy consumed by a control group of homes. Though such estimation suits the utility's needs in satisfying regulatory requirements that across large portfolios of homes energy savings does in fact occur, it does not and is not intended to quantify the energy saved by improving any given home.

What about the homeowners themselves? It would seem that homeowners would have the most interest in knowing how much energy has been saved by improvements they have made. In fact, homeowners have no need to quantify energy savings as a result of improvements on their homes because the knowledge itself has no impact on them. Once a home improvement is made, the cost of the improvement is sunk and all energy savings as a result of the improvement accrue to the homeowner in the form of lower utility bills. As there is no need to divide the saved energy value accrued, there is no need to quantify energy savings, and the homeowner is content receiving lower utility bills without needing to know specifically how much lower the bills are than what they would have been had the home improvement not been made.

By analogy, the owner of a car with a hybrid electric engine does not typically do the math on mileage driven and gasoline prices after the fact to see whether the savings on fuel use over time compensates for the increased price of the hybrid engine. The reason is because the decision to buy the hybrid engine has already been made, and the cost of the car is now a sunk cost. As all of the value of reduced gasoline costs accrues to the car owner, it is sufficient for the car owner to know that he or she is saving money by going to the pump less frequently.

If, on the other hand, a financial entity were to offer to pay for a portion of the upgraded engine up front in exchange for a portion of the fuel savings over time, a contract would have to be in place between the car owner and the financing entity over how exactly to determine the counterfactual of how much would have been spent on gasoline had the hybrid engine not been bought, thereby determining how much the car owner would owe the financial entity over time.

Without a financial imperative to divide the accrued value of saved energy, there is no reason to establish a defendable counterfactual, and so the car owner goes along his or her way knowing that he or she is saving money but not knowing exactly how much, and therefore whether in fact the hybrid engine is worth the investment. Similarly, quantifying the value of energy savings from home improvements is only required when the value of the saved energy accrues to multiple parties, which it rarely does.

Measuring energy savings, as it turns out, is a form of collective action problem: Each individual homeowner benefits by having a better understanding of how much energy other homeowners are saving and how the savings were achieved. The homeowner attempts to make rational economic decisions to invest up front in home improvements with an estimate of the savings over time, but as the homeowner accrues all value for better or worse, the homeowner has no incentive to check in hindsight whether the savings estimates came to fruition (and may in fact have a cognitive bias against knowing the truth lest the truth reveal that realized savings are in fact less than believed or expected). As a consequence, we fail to socialize the true impact of our home improvement decisions, thereby depriving the next homeowner attempting to make a rational economic decision of critical information.

Understanding how much energy others have saved would give us all guidance into which contractors to choose, which home improvement projects to undertake, which products to buy, which programs to participate in, and how much money to spend in order to optimize our own returns on investment. Such information could unlock currently unavailable financing for home energy improvement projects from lenders who would be willing to invest in home energy efficiency if assurance were available that savings will be realized on a portfolio of homes. But for each of us as individuals, we have no incentive to know how much we are saving, and so together as a group we lack the information required to make individually rational economic decisions regarding energy efficiency. Ironically, we would all like to how much has been saved from others' home improvements, but how much our own home improvements are saving us does not interest us enough to find out.

Finally, what about financial entities? Any financial entity that invests in energy efficiency where the return on investment is dependent on energy savings must by definition measure the energy saved. The fact that measuring home energy savings is so rare indicates that external financing for home energy improvements on the basis of saved energy is also rare. More common is the availability of home energy improvement loans, either as unsecured loans or in addition to the home's mortgage, though in neither case is energy savings measured over time, as the homeowner is required to pay back the loan at the same rate regardless of how much energy is in fact saved.

Financing for Energy Savings

Home energy improvement loans provide a source of capital to homeowners to finance home improvements while simultaneously providing an investment opportunity for external parties. Examples of home energy improvement loans include utility on-bill financing and credit lines extended by credit unions. A property-assessed clean energy (PACE) bond is a loan extended to homeowners, usually by municipalities, to finance home improvements with the intention of saving energy, with a senior lien placed on the mortgage of the home for the value of the loan (What Is PACE? n.d.). Though this lien solves problems for the bond issuer, such as collateralizing the loan and securing a payback mechanism, mortgage lenders and securitizers including Fannie Mae and Freddie Mac have raised objections to PACE bonds on the grounds that the lien subordinates the underlying mortgage, thereby reducing its creditworthiness. While this objection is somewhat technical in nature, it highlights the underlying issue of whether home energy improvements are in fact best collateralized against the mortgage, against a future expectation of energy saved, or unsecured to the homeowner directly. Each financing option would carry a different risk profile to the investor and thus would command different rates of return in order for the investment capital to be unlocked.

What all of these current outside financing plans for home energy improvements have in common is that the repayment is not tied to energy savings, and therefore actual energy savings are not typically quantified. Tying loans to energy savings poses challenges to the lenders, such as the increased variability of the return on the investment, quantifying the energy savings itself, collecting payment from the homeowner on the basis of that savings, and explaining the underlying rationale behind the variable amounts to participating homeowners.

Not quantifying energy savings is a point of convenience for lenders, but is not without cost to homeowners. Without assurance that energy is in fact being saved, homeowners risk that improvements do not provide a return on investment to them that justifies the loan. If in fact this is the case, unless the improvements disproportionately add to the home's resale value, home energy improvement loans disconnected from actual energy savings simply increase a homeowner's debt burden without any actual financial benefit to the homeowner.

The underlying reason why energy savings achieved through home energy improvement loans are not quantified is thus that all energy savings value currently accrues to the homeowner alone, and the homeowner is not particularly interested in a precise determination of past energy savings; knowledge of the actual amount does not impact future behavioral or capital allocation decisions. Homeowners all have an economic interest in saving energy, and therefore money, but no specific economic interest in knowing how much they would have saved had they taken different action in the past.

Another mechanism for financing a home improvement would involve an outside entity's paying for and owning the physical improvement, though such a proposition is highly problematic. Regulated utilities owning assets in a home does have precedent—telecommunications utilities regularly own the TV cable box and charge the homeowner a fee per month for usage. The electric utility already owns the electricity meter typically located on the side of the house and may also own a direct load control device that remotely interrupts power to appliances such as an air conditioning compressor, water heater, or pool pump. However, these externally owned devices are directly related to providing the regulated service: the cable box is required to receive television channels, and the utility meter is required to appropriately charge for the commodity usage. Homeowners are typically willing to accept the utility's deploying a direct load control device on their home if it enables them to sell unneeded or unwanted peak load power back to the utility.

By extension, an electric utility or other outside entity could own the HVAC system in a home. External investors do sometimes own assets within commercial buildings where an energy service contract is in place. In some cases, this equipment in the mechanical room may have a physical barrier such as a fence around it to demark the separate ownership.

However, owning a home HVAC system is quite different from owning a meter or cable box, as the output of the HVAC system—hot or cold and dehumidified air—is not a product or service that an electric utility sells. Electric utilities do not generally have the capability or desire to put on their own balance sheets a physical asset within a home that is subject to being tampered with by the homeowner. In any event, an electric utility owning a home HVAC system and simultaneously charging for the input while not being accountable for the output presents an inherent conflict of interest. The burden of ownership would need to fall on a hypothetical energy services provider.

From Energy to Energy Services

Energy efficiency in any context is the ratio of desired energy services output to required energy commodity input. Homeowners do not in point of fact want electricity and natural gas; homeowners want cold milk, cool dry indoor air, hours of television viewing, and other services that energy inputs provide.

The core problem in structuring a functioning economic system around energy efficiency comes down to the conflict of interest of the entity selling the commodity, be it electricity, natural gas, or another fuel. The commodity provider, whether a regulated investor-owned utility, public utility, or retail energy provider, has an economic interest in the residential customer (or rate payer to the regulated utility) consuming more of the commodity. Each type of energy commodity provider is subject to constraints that artificially counteract this economic interest, but none of these constraints involves quantifying energy savings at the utility customer level.

The ultimate way to overcome the inherent contradiction of the commodity input provider attempting to sell energy services output is for an entity to offer and profit from the efficient delivery of the output energy services rather than the input commodity. For example, consider an air conditioning services company that sends a homeowner a bill for cool, dehumidified air. This company would be responsible for owning the compressor and air handler at the home, maintaining this equipment, and paying the portions of the electric bill that the air conditioning system consumes. This company would be very aware of the energy savings value of work that both has been done and could be done at the home. The company's profitability would be directly tied to its ability to deliver to the homeowner the most efficiently conditioned air that it can, subject to its own economic constraints. As a result, the company would be continuously making capital allocation decisions on the basis of which air conditioners to fix or replace for which homes based upon projected returns in energy saved. This cold air provider would focus on providing the most efficient cooling services to as many homes as possible, thereby removing the accountability for air conditioning efficiency from the homeowner and institutionalizing it in a managed process.

A home energy services company is a hypothetical entity that would deliver energy services rather than energy inputs. An energy services provider would, in theory, own and service (or arrange for the service of) energy consuming appliances and devices such as the HVAC system, lightbulbs, and possibly even appliances such as the refrigerator and clothes dryer, somewhat like the landlord of a furnished home would but without owning the home itself. Instead of the homeowner's buying energy inputs such as electricity and natural gas, the homeowner would buy cold or hot air from the HVAC system, cubic feet-hours of refrigeration, and even lumens of light from fixtures. The energy service company would be fiscally responsible for ensuring that the energy services were delivered in the most cost-effective way.

The home energy services provider would have an imperative to quantify energy savings not only of homes, but also of services provided. The energy services provider would be required to make capital allocation decisions across its portfolio of customers and throughout its balance sheet of assets on the basis of return on investment of its portfolio—appliances would receive repairs and service when it makes economic sense to provide such rather than on a regular schedule, and would be replaced at the company's discretion.

Critically, the energy services provider would have the ability to select the homes to provide services to on the basis of which homes would have the greatest potential to return capital on energy saving improvements. The energy services provider would view homes as reservoirs of energy efficiency waiting to be monetized, much in the same way that geologists inform drillers where and how to drill oil wells in order to most cost-effectively produce oil. Only by considering each home individually based upon the energy efficiency potential of that specific home can capital be efficiently allocated to produce home energy efficiency as an economic resource.

For numerous reasons, the home energy services provider as described above will remain a hypothetical entity. Aside from the challenges of metering energy service quantities such as HVAC system output, a homeowner-level energy services model would require a level of coordination and critical mass that would be very difficult to justify with the margins that the home energy services provider could reasonably expect to receive absent sufficiently high energy prices. Owning and amortizing corporate assets physically located within homes would be a difficult financial proposition from which to raise debt in order to finance the business model. The model itself would strike many homeowners as confusing and complicated, particularly as the energy services provider would be responsible for some, but not all, of a home's energy costs.

Short of a hypothetical energy services provider, there are two other models in which quantifying energy savings plays a role. The first is the traditional energy service company as currently implemented by building owners, extended into the residential market. Energy service companies offer energy savings performance contracts to energy consumers, in which the energy service company pays for or arranges payment for infrastructure improvements as part of building retrofits. The building's management signs a long-term agreement with the energy service company, some up to 25 years, to pay the value of energy saved over time in order to benefit up front from the improvements made to the building, with a detailed agreement in place over how energy savings will be determined and money repaid. The energy savings performance contract is thus a form of bond secured by the value of the saved energy over time.

The energy service company model is most commonly used to pay for improvements made to aging government buildings; most performance contracts are delivered to buildings within the "MUSH" market (municipal and state government, universities, schools, and hospitals). The MUSH market lends itself well to the energy savings performance contract model for multiple reasons, including long-term tenancy, the longevity of the buildings, the size of the contracts, incentives provided by governments to external parties to pay for building retrofits, experience with the building types, established contract terms, and a track record by energy service companies of activity in this market. Also important is that the long-term nature of the contracts allows for a lower hurdle rate. Privately owned commercial and industrial facilities generally require rapid payback on the order of a few years or less to justify a project, and if the energy savings project clears the company's hurdle rate, the project would likely be self-financed rather than financed through an energy savings performance contract.

For a residential energy service company to be profitable, the company would need to overcome additional challenges specific to the energy savings performance contract model. The homeowner would need to be billed for the energy savings, and the payment would need to be collected, either as a line item on an existing bill (e.g., a municipal or utility bill), or as an additional bill to the homeowner. A standard M&V methodology would need to be agreed upon by all parties and be explainable to the homeowners should they inquire how exactly the amount they are paying for energy efficiency services is determined.

Though such challenges are significant, they are not impossible to overcome, and it is quite reasonable to expect that a residential energy service company could emerge. Essentially acting as a home energy efficiency developer, the company would offer only the limited range of services that clear the company's internally determined investment hurdle rate, likely including sealing air leaks, fixing ducts, adding missing insulation, and other low marginal-cost work on pre-identified energy inefficient homes.

The other model in which quantifying energy savings plays a role involves fixed utility bills. Some utilities offer fixed bills, in which homeowners pay a predetermined fixed amount per month. Such programs are generally not popular because the average payment over time tends to be higher than within traditional programs that charge for actual use, as utilities must factor in the risk of the homeowner's overconsumption. Also, offering the homeowner a fixed payment gives the homeowner a perverse incentive to overconsume, much like the tendency to overeat at an all-you-can-eat buffet.

In addition to utility fixed bill programs, some private companies also offer homeowners fixed bills. Such companies evaluate homeowner energy consumption through an actuarial exercise in which the majority of participating homeowners underconsume while a small percentage overconsume relative to the fixed payment, much as insurance companies adjust for the different utilization by their customers of the insurance plans they offer. These companies have the option and financial incentive to find ways of helping their customers save energy, and thus are attuned to both the energy-saving potential within their customers' homes and also the value of energy-saving measures initiated by the private company itself. Regardless, the company does not have a great incentive to invest in its customers' homes' energy efficiency as (1) the customers are able to drop the service shortly after the investments in their homes are made, and (2) the company is able to unsubscribe those customers who perpetually overconsume. In this model, quantifying energy savings can be important, but doing so does not necessarily lead to large investments in home improvements that save energy.

A possible hybrid model would be a cross between an energy services company and a fixed utility bill provider, in effect a turnkey energy services provider. This entity would own and service home appliances and also be responsible for paying a homeowner's utility bill, all for a monthly fee, thereby capturing the value from appliance improvements for itself. Though theoretically possible, the entity would be subject to excessive risk from homeowners' not taking care of the appliances in the home and overconsuming energy. Removing unprofitable homeowners from the program would involve one of several unpalatable recourse options, such as a forced sale of the assets to the homeowner through the mortgage, physically entering the home and removing the appliances, or alternatively remotely shutting appliances off until payment is resumed. As a result of these potential complications, such a model is unlikely to arise any time soon.

To summarize, quantifying energy savings is a requirement to unlock outside capital in home improvements that are tied to energy savings, but delivering a service that saves energy and then recapturing the energy savings from the homeowner as an energy efficiency charge is currently prohibitively complicated. The only remaining option for dividing the energy savings value accrual involves monetizing the energy savings from home improvements while in parallel allowing the actual value of the commodity not consumed to accrue to the homeowner. Two such monetization models that fit this description are currently available.

Energy Savings Certificates and Carbon Credits

Widespread measurement of energy savings will only occur when the savings themselves, when properly measured, have monetary value. In some instances, energy savings are tradable as energy savings certificates (ESCs), or "white certificates," which can be bought by utilities, corporations, or others in order to offset their own energy use or comply with regulated portfolio standards pertaining to energy efficiency (Energy Saving Certificates, 2008). ESCs monetize energy efficiency project savings in the same way that renewable energy credits monetize the output of renewable energy systems such as solar panels and wind turbines. An ESC is typically expressed in units of electrical (MWh) or thermal (Btu) energy conserved on an annual basis, with one ESC representing 1 MWh of energy saved. The ability to sell ESCs provides homeowners and third parties additional incentive to invest in energy savings projects.

Though most states have energy efficiency resource or portfolio standards with targets on energy efficiency development, only four states (Connecticut, Pennsylvania, New Jersey, and Nevada) have specific provisions in which third parties generate ESCs and sell them to utilities that are seeking to comply with energy efficiency targets, typically from commercial and industrial improvements (US Environmental Protection Agency, 2014b). In New York, the New York State Energy Research and Development Authority (NYSERDA) has piloted a voluntary ESC market based on M&V standards from an existing utility energy efficiency program. NYSERDA aggregates ESCs, auctions them, and then uses the proceeds to fund public programs for energy efficiency. The use of existing programs and M&V protocols helps to provide credibility for the ESCs generated as well as lowers costs associated with the ESC generation program. Regional transmission organizations that operate wholesale energy markets, such as PJM and ISO New England, accept some commercial energy efficiency as a resource in forward capacity markets. Internationally, Australia, United Kingdom, France, and Italy also have ESC trading or procurement systems in place (Friedman et al., 2009).

Though aggregation of home improvements is not generally referenced as an independently procurable ESC resource option, renewable and energy efficiency portfolio standards are open enough in their language to account for multiple sources of energy efficiency or renewable energy that may prove cost-effective to develop in order for a state to meet its resource targets. A portfolio of home improvements with savings proven by compliant M&V could certainly serve as a valuable resource option to utilities, particularly in areas with high peak load procurement costs from the wholesale power market or bulk power provider, and also in areas where new electricity generation or distribution infrastructure is not a cost-effective option. A wider use of ESCs opens the possibility of enabling market forces to drive effort and capital toward developing the most cost-effective energy savings opportunities available, including for home improvements (Friedman et al., 2009).

Carbon credits are similar to ESCs in that they are a mechanism to monetize energy savings. Carbon credits are created when an entity acts to reduce emissions of carbon dioxide or other greenhouse gases, with one carbon credit representing one ton of carbon dioxide not emitted. Carbon credits are typically purchased by an entity seeking the right to emit carbon dioxide as part of an industrial or power generation process. Though markets for carbon credits are not well developed, some companies and entities voluntarily procure carbon credits in order to offset their own emissions, and carbon credits are actively registered and traded on the European and Chicago climate exchanges.

Determining carbon savings from home improvements is reasonably straightforward if proper quantification of energy savings is performed, such as through the use of whole-building energy use modeling. For natural gas savings, each therm not consumed results in 11.7 pounds of carbon dioxide not emitted. For electricity, the calculation of carbon dioxide emissions per kilowatt-hour is subject to the mix of electricity-generating sources (e.g., coal, gas, nuclear, renewable) providing electricity at any given time to the distribution grid to which a home is connected. However, emissions proxies are available from the Emissions & Generation Resource Integrated Database (eGRID), published by the US Environmental Protection Agency (EPA, 2014a), which shows the total carbon dioxide emissions per kilowatt-hour by US subregion. By converting quantified home energy savings to carbon credits and selling those credits to a willing buyer as an aggregated bundle, carbon credits could be leveraged in parallel to ESCs as an additional source of capital to offset the cost of home improvements.

How Utilities Can Help

Because utilities store one of the key pieces of information required to compute energy savings (i.e., metered quantity), utilities can make measurement of energy savings easier for any interested party, homeowners or otherwise, by making historic usage data easier to collect and analyze.

Nearly all electricity and natural gas utilities already provide online access to the historic energy usage of a home for the homeowner to view. The primary purposes of providing this information are for bill collection, reconciliation, and dispute resolution, but as explained above this very same usage information serves the useful purpose of being an input into energy savings determination.

The most important numbers available from a utility that are needed to measure energy savings are the meter read date (and if available, time) so that the weather during the period can be determined, and the amount of commodity consumed, typically in units of kWh of electricity and therms of natural gas used. Also, the total charge, or a combination of fixed charge and per unit charge, are highly useful in converting from units consumed to amount billed and therefore dollars saved. These three numbers (meter read date, usage, and total charge) should ideally be easily portable; any homeowner should be able to copy and paste this information into a spreadsheet or download it as a spreadsheet should they wish to do so.

Other formats to easily retrieve data are also available. A link that would enable the download of a CSV (comma-separated values) file would also be useful as such files can be opened by a range of software programs including spreadsheets. Some utilities with data of higher resolution than monthly bills, such as daily or hourly consumption, use a "green button" which enables download of an XML (extensible markup language) file (XML is a machine-readable file format and thus is helpful for data analysis). A copy-and-pasteable data table and/or a spreadsheet or CSV file download would be a good minimum standard of data accessibility to enable any homeowner or contractor to perform home consumption data analysis.

Either intentionally or accidentally, many utilities have made it difficult for homeowners to access these data. Some utilities go no further than having downloadable bill images available one at a time, which can take a homeowner quite a while and many mouse clicks to access and then record all required values. Having to do so just to measure energy savings should be completely unnecessary. To assess the ease or difficulty of analyzing a given utility's consumption data provided to the homeowner through an online portal, the following simple test could be applied.

The Building Performance Institute (BPI), a standards-setting organization for home performance professionals, developed ANSI BSR/BPI-1200-S-2015, "Standard Practice for Basic Analysis of Buildings," to be used by home contractors in evaluating the energy performance of homes they inspect (Building Performance Institute, 2015). Contained within this standard (p. 31) is the following equation for estimating a home's baseload (i.e., non-temperature-related) energy use:

Annual baseload = $12 \times 1.1 \times$

(average of the three lowest months' usage among the last 12 months)

To find the annual above-baseload usage, subtract the annual baseload from the sum total of the past 12 months' usage.

The process of retrieving the last 12 months of usage data into a spreadsheet to perform this basic math should take no longer than a minute following log-in to the online portal. If it takes any longer than a minute, the utility has erected an unnecessary roadblock to homeowners' use of their own data for the purposes of analyzing their own usage, or enabling a home contractor to do so on the homeowner's behalf.

In addition to simplifying data access, utilities could also help by providing more historic data in homeowners' online accounts. Utilities generally keep only two years of historical consumption, and some keep only one. With only two years of data, a retrofit that occurred more than one year ago will not have enough data available for the determination of pre-retrofit baseline usage in a whole-building energy use model. All utilities could certainly help make it easier to measure energy savings by following the lead of some electric cooperatives and making all prior utility bills for a given homeowner at the same residence easily available through the homeowner's online account; at a minimum, three years of historical bills would be a reasonable guideline. With three years of data available, the savings from home improvements performed up to two years earlier could be easily quantified.

The Future of Quantifying Home Energy Savings

Though ASHRAE-14 provides a technical framework for the measurement of energy savings, putting it into practice is left to the practitioner. Energy engineers and M&V professionals may have a preference (or may have clients who have a preference) for one set of assumptions over another—for example, whether to use average temperature or degree days for weather correction, CV(RMSE) or R^2 for statistical evaluation, whether or not to vary in the model the heating and cooling balance point temperatures of the home, and discretion over which points to remove as outliers, are all subject to the preferences of the professional evaluator. Although NREL's Uniform Methods Project usefully updates and complements ASHRAE-14, the Uniform Methods Project is provided as an option to those seeking generally accepted practices but is not intended to standardize practices across the industry as the only manner in which savings can be reliably determined.

For value to accrue to parties other than the homeowner, the techniques employed for measuring energy savings need to comply as a minimum standard with those imposed by the provider of the value. In the case of a regulated utility buying ESCs, the means of determining the energy savings must comply with regulatory statute regarding the determination of energy savings under the state's existing M&V statutes—ASHRAE-14 and/or the Uniform Methods Project are certainly acceptable under these standards, as they generally provide sufficiently rigorous savings determination when properly followed. However, other parties may require more stringent, specific, and—most importantly—standardized savings measurement techniques.

To unlock broader capital markets to home energy efficiency as an investable asset class, a standard method for energy savings measurement may be required. Otherwise, the owners of the energy efficiency investments will not have sufficient confidence in the valuation of their investments to be able to treat them as investment grade assets. Further, if energy savings measurement is taken out of the hands of industry professionals and put into the hands of those serving independent investors, whoever is measuring the energy savings has an incentive to overestimate the quantity of energy savings through questionable statistical techniques. Such behavior, if it were to emerge, would result in a black mark on the home energy efficiency industry. An increase in reliance on energy savings measurement may require a corresponding increase in the standardization of energy savings measurement techniques. As energy savings is a computed rather than metered quantity, increasing standardization of measurement techniques should in theory lead to more trust in the realization of actual energy savings and therefore lead to more investment in home improvements.

One possible improvement would be a common standard for energy efficiency measurement. In March 2015, BPI announced a proposal to develop a new standard to be called "Protocol for Quantifying Energy Efficiency Savings in Residential Buildings," a standard to be cosponsored by the Air Conditioning Contractors of America, with the purpose of creating a unified methodology to commoditize and consistently value residential energy savings at the portfolio level. Taking the idea of standardizing energy savings measurement one step further, a group called Open Energy Efficiency Meter is developing open standard source code, called the EE Meter, for the calculation of energy savings (Open EE Meter, 2015). The group aspires to standardize the metric used to account for energy savings through a publicly available algorithm.

Another possible future scenario for energy savings measurement involves high-resolution retrofit isolation. As the number of communicating devices in our homes proliferates, computers will increasingly have the intelligence to know our homes' operating conditions. Low-cost sensors, two-way communicating devices, and cloud computing together empower a home that is self-aware in terms of its ability to efficiently consume energy. Whirlpool, GE, and Samsung today market communicating appliances; likewise, smart thermostats from companies such as Nest, Honeywell, and Trane already have the capability to improve the operation of homes' climate control systems.

One day we will not only have smart thermostats, but likely also smart HVAC systems, plumbing systems, and natural gas systems—and maybe even smart windows and walls through embedded sensors or through computer modeling of their condition via data such as indoor and outdoor climate monitors in what would in essence be a perpetually (or periodically) updated calibrated simulation. One day there might be home monitoring services that monitor energy efficiency as well as livability indicators such as allergens, pathogens, and pollutants in a similar manner as how security companies monitor our homes' alarm systems. Such a home monitoring company would be keenly aware in real time of our home's energy efficiency. This company would be ideally suited to provide homeowners with custom advice on improving their homes and also to quantify for homeowners the value of improvements.

The near-term future of energy savings measurement will most likely involve the slow but steady increase in the use of utility meter data to calibrate energy savings estimates. M&V professionals can use utility bills to recalibrate estimates of home energy savings after the fact, but this practice is not universal and is not typically performed outside of the context of utility-driven programs. As it becomes easier to digitally access utility billing data, to import the parameters determined from building models generated by contractors into energy analysis software, and to access high-resolution weather data, we will likely see an increase in whole-building energy use modeling of homes beyond the utility-driven context, even without further standardization of techniques for energy savings measurement.

Who will be the early adopters of new practices for per-home energy savings measurement over time? The most likely groups are those who currently measure energy savings and seek to expand doing so through simpler and cheaper means. One such group includes research institutions in the field of building science contracted by utilities, government agencies (such institutions themselves may be parts of government agencies), and private companies to examine the impact of different building materials or methods on home energy use. Their work often includes the placement of additional data collection devices, such as thermometers and hygrometers, in various places throughout the tested homes, as well as the detailed analysis of all available relevant data. Though these processes return high-quality results, they are prohibitively costly and excessive for non-test homes. Through quantifying energy savings across many more homes, an expansion of simple home energy savings methodology would increase our understanding of building science as it applies to energy use in modern homes across climate zones, demographics, architectural elements, and other interacting domains.

In addition to building science researchers, another early adopter group is likely to be university-driven home energy efficiency programs (or other home energy programs with informational objectives). Such programs are funded through public or other research grants to improve home energy efficiency. They usually serve the dual purpose of making participating homes more comfortable and energy efficient while also providing insights on energy economics, building science, program management best practices, and other aspects of home improvement with relevance beyond the programs themselves. These programs have a strong impetus to quantify energy savings over time, including for example accurately reporting program impacts to grantors and other stakeholders, improving program outcomes through continuous homeowner engagement and contractor evaluation, and increasing the value and richness of lessons learned. Some of these programs already measure energy savings in participating homes, and it is likely that as best practices for doing so are shared, similar programs will follow their lead. Two such programs are the E-Conservation Program and the Duke Carbon Offsets Initiative, which use ResiSpeak for utility bill collection and whole-building energy use modeling to determine energy saved for each participating home.

Conclusion

Though methodologies for quantifying home energy savings are well established, today nearly no one systematically quantifies energy savings from home improvements. The primary reason is because no party other than the homeowner directly benefits from the actual energy saved following a home improvement. As all of the energy savings accrue to the homeowner, the homeowner also has no specific need to know how much energy in fact was saved. Though widespread quantification of energy savings from home improvements would provide numerous societal benefits as a result of the accountability and transparency derived, home energy savings will continue to go unquantified until a business model is established and popularized in which an outside party directly benefits from the energy savings that result from the energy efficiency improvements made to a home.

Energy savings certificates and carbon credits are two vehicles to monetize energy savings from home improvements that require quantification of energy savings to value, but these vehicles are not yet widely used. Quantifying energy savings will become easier over time as data become more available and standardized, but until business models emerge that provide sources of capital for home improvements that are returned to parties other than the homeowner on the basis of energy saved over time, quantification of energy savings from home improvements will continue to be the exception rather than the norm.

Chapter References

Agnew, K., & Goldberg, M. (2013). Chapter 8: Whole-building retrofit with consumption data analysis evaluation protocol. In T. Jayaweera & H. Haeri (Eds.), *The uniform methods project: Methods for determining energy efficiency savings for specific measures*. National Renewable Energy Laboratory subcontract report NREL/SR-7A30-53827. Retrieved from http://energy.gov/sites/prod/files/2013/11/f5/53827-8.pdf

American Society of Heating, Refrigerating and Air-Conditioning Engineers, Inc. (AHSRAE). (2002). *ASHRAE Guideline 14-2002: Measurement of energy and demand savings*. Retrieved from https://gaia.lbl.gov/people/ryin/public/Ashrae_guideline14-2002_Measurement%20of%20Energy%20and%20Demand%20Saving%20.pdf

Building Performance Institute, Inc. (2015). *Standard practice for basic analysis of buildings*, ANSI BSR/BPI-1200-S-201x. Retrieved from BPI Standards: http://www.bpi.org/files/pdf/ANSI%20BPI-1200-S-2015%20Standard%20Practice%20for%20Basic%20Analysis%20of%20Buildings.pdf

Doty, S., & Turner, W. C. (2012). *Energy management handbook* (8th ed.). Lilburn, GA: Fairmont Press.

Energy Saving Certificates. (2008). *The bottom line on...: Answers to frequently asked questions about climate and energy policy*. Issue 10. Washington, DC: World Resources Institute. Retrieved from http://www.wri.org/sites/default/files/pdf/bottom_line_energy_savings_certificates.pdf

Energy Upgrade California. (2015). Gas savings realization by contractor [Table]. Retrieved from www.caltrack.org/caltrack.html

Friedman, B., Bird, L., & Barbose, G. (2009). *Energy savings certificate markets: Opportunities and implementation barriers* (pp. 1–10). Conference paper NREL/CP-6A2-45970. Presented at the American Society of Mechanical Engineers (ASME) Third International Conference on Energy Sustainability, San Francisco, CA.

International Performance Measurement & Verification Protocol Committee. (2002). *Concepts and options for determining energy and water savings* (Vol. I). National Renewable Energy Laboratory report DOE/GO-102002-1554. Retrieved from http://www.nrel.gov/docs/fy02osti/31505.pdf

Jayaweera, T., & Haeri, H. (2013). *The uniform methods project: Methods for determining energy efficiency savings for specific measures.* National Renewable Energy Laboratory subcontract report NREL/SR-7A30-53827. Retrieved from http://energy.gov/sites/prod/files/2013/07/f2/53827_complete.pdf

Open EE meter [Homepage]. (2015). Retrieved from www.openmeter.org

US Environmental Protection Agency. (2014a). *eGrid* (9th ed. with year 2010 data; Version 1.0). Retrieved September 25, 2015, from EPA Clean Energy Resources, http://www.epa.gov/cleanenergy/energy-resources/egrid/

US Environmental Protection Agency. (2014b). *Survey of existing state policies and programs that reduce power sector CO_2 emissions. Appendix for state plan considerations: Technical support document* (ID No. EPA-HQ-OAR-2013-0602). 46 pp. Retrieved from http://www2/epa/gov/

What Is PACE? [webpage]. (n.d.). Retrieved from www.pacenation.us/about-pace/

Considering the Effect of Incorporating Home Energy Performance Ratings Into Real Estate Listings

Gabrielle Wong-Parodi, Jim Kirby, Ryan Miller,
and Melanie Girard

Introduction

Increasing the overall energy efficiency of the US housing stock depends in part on consumers' willingness to purchase energy efficient homes. Indeed, research efforts have sought to identify, analyze, and understand this "green consumer" across a wide range of disciplines, from marketing and economics to industrial ecology to sociology and psychology. Peattie (2010) describes green consumption as being strongly influenced "by consumer values, norms, [and] habits," but that it is also "complex, diverse, and context-dependent." In this chapter, we first focus on consumers' behavior as it relates to real estate purchase decision-making. We then review the literature on consumer home buying behavior in general and for green consumption in particular, with a discussion of how and why people search for information when weighing home options. Next, we present literature on the influence of energy efficiency and related rating systems on home purchasing decisions, including a discussion on best design practices for rating systems that help consumers make informed decisions. Finally, the chapter concludes with a brief overview of the two dominant rating systems (the Home Energy Rating System Index and Energy Star) in the US, their influence on purchasing decisions, and recommendations for best practices for evaluating such systems.

Home Purchasing Behavior

Theoretical Background

Purchasing a home is one of the most important and challenging decisions that consumers make. Strategic decisions (Bazerman, 2001; Grønhaug et al., 1987) such as the purchase of a home or investment decisions (Henry, 2005) are marked by consumers' high involvement in the process, long-term commitment of resources, and truncated budget for other goods and services (Arndt 1976; Grønhaug et al.,1987; Grewal et al., 2004; Rosenthal, 1997); Kos Koklič & Vida, 2009). Thus, this class of decision involves several important categories of decisions (Kos Koklič & Vida, 2009), including allocation of household economic resources (e.g., travel, restaurants), categorization of choosing between various product groups (e.g., apartments or houses), defining a more narrow product category (e.g., houses of a given size), and decision making within the defined product category.

When we consider the context in which a homebuyer is making a decision to purchase a home, a good starting point is the empirical literature on durable goods. Indeed, the house is arguably the most important durable good in the household (Hempel & Punj, 1999) and many studies of consumer decision making indicate there are many similarities among the buying processes for various durable goods (Bayus, 1991; Cripps & Meyer 1994; Grewal et al., 2004; Hauser & Urban 1986; McQuinston, 1989; Punj & Brookes, 2002). This literature suggests that compared with buying convenience products, consumers perceive large ticket purchases as risky, sometimes even as traumatic (Chaudhuri & Holbrook, 2001; Mitchell, 1999). In short, durable goods purchasing decisions are complex, especially when consumers perceive the price to be high (Kos Koklič & Vida, 2009).

Purchasing a home is a complex decision, as the market offers a rich variety of price and quality, so it is likely difficult for consumers to specifically apply previous knowledge in their decision making process. In this context of high involvement and complexity, consumers must consider several possibilities, compare them, and ultimately make a selection. Given this context, we now turn to a conceptual model of the home purchase decision-making process.

Conceptual Model

Kos Koklič and Vida (2009) describe a useful conceptual model comprising Peter and Olson's (2002) cognitive processing model and Hawkins and colleagues' (2003) general consumer behavior model. The former describes a model where the "consumer decision-making process is a goal-directed, problem-solving process" (p. 168). The latter depicts both internal and external factors that contribute to consumers' self-concept and lifestyle, and thus influencing decision making. This is not a static process, and life experiences and purchases made update the internal and external influences. Thus, a home should fulfill a homebuyer's goals and needs, and at the same time the home chosen will reflect a person's self-concept and lifestyle.

According to this model, lifestyle and self-concept influence a homebuyer's needs, preferences, and desires. This in turn influences goals and, subsequently, the homebuyer's decision with respect to the home in question. Several internal factors play into a homebuyer's lifestyle and self-concept, such as involvement, feelings, experience, knowledge, motivation, and personality. There are also as a number of external factors such as culture, subculture, reference groups (such as those that set social norms, etc.), family, social class (e.g., socio-economic status), and demography. Indeed, this model suggests that a person's lifestyle and the meaning a person imbues into homeownership influence the homebuyer's needs and desires concerning the home itself. People also need to incorporate new information in the decision-making process.

Given the complex nature of home buying, in the early stages, purchasers usually do not have sufficient information. Thus as new information is acquired, desires and goals for the home are adapted accordingly. In the next section, we turn our attention to how homebuyers search for information.

Information Search

As Srinivasan (1990) has noted, there are at least three main theoretical approaches toward consumer information search that we can consider: economic explanations, psychological explanations, and information processing explanations. Below, we summarize Lin and Lee's (2002) overview of these three approaches.

Economic. The economic approach posits that information search is conducted in terms of the search's costs and benefits. Here, maximizing utility and imperfect information are two important assumptions wherein rational consumers try to maximize their utility, but they do not know about all of the prices at a given time due to constant changes in market prices. Therefore, they tend to search among a variety of sellers for a good price when a range of prices is available.

According to this theory, the consumer finds that increased searching results in diminishing returns, which is expressed by reduced expected minimum price. Thus, the amount of search that a consumer will engage in is determined by the marginal cost of the search and its marginal return or benefit (Butters, 1977; Kohn & Shavell, 1974; Ratchford, 1982; Rothschild,

1973; Salop, 1976; Stiglitz, 1979; Telser, 1973; Weitzman, 1979). The search for information is negatively related to its costs and positively related to its benefits.

Here we can think of costs as what a consumer must sacrifice in order to obtain and process information (Bloom, 1989; Russo, 1988; Russo & Leclerc, 1991)—both direct and indirect costs. The direct costs of search include monetary spending, time sacrificed, and physical and psychological effort (Bettman, 1979). The indirect cost can be thought of as the opportunity cost or the lost utility of alternative uses of time spent searching. Research has also found that the perceived costs of search negatively affects people's motivation to search (Bettman, 1979; Farley, 1964; Stigler, 1961). Moreover, studies have found that as information search decreased, costs increased (Srinivasan, 1987). Indeed, there are many search costs associated with purchasing a home. These include the time spent researching and viewing homes and talking to real estate agents and mortgage brokers.

However, just as there are costs to searching, there are also benefits. Here, benefits can be thought of as the value of achieving one's desired goals (Gutman, 1982; Olshavsky & Wymer, 1995). Several benefits of information search include obtaining a lower price, higher quality, increased satisfaction with the decision, and (perceived) reduction of risk (Bettman, 1979; Punj & Staelin, 1983; Bennett & Harrel, 1975; Howard & Sheth, 1969).

Psychological. A psychological approach suggests that a number of factors influence information search behavior, including individual characteristics (e.g., personality traits), the type of product (e.g., home, food), and other search-related variables (e.g., time and resources available). Howard and Sheth's (1969) motivational approach is useful in this regard, as it suggests that attention—regulated by the stimulus ambiguity-arousal relationship—initially motivates search behavior.

In the context of buying a home, a psychological explanation for search would point to motivation for buyers to engage in search activities (Burnkrant, 1976; Hansen, 1972; Howard, 1977; Howard & Sheth, 1969; Nicosia, 1966). Here motivation can come from a buyer's goal orientation, which differs between people who are "optimizers" and those who are "satisfiers" (Srinivasan, 1990; Swan, 1969; Wright, 1975). Optimizers engage in a more thorough search than satisfiers do. Motivation can also come from the degree of involvement (e.g., such as long-term economic investment), which affects perceived importance (Bloch & Richins, 1983). The greater the involvement a buyer has in the home purchasing process, the greater the motivation to search for information (Beatty & Smith, 1987; Schmidt & Spreng, 1996). Finally, the perceived costs and benefits of information search also serve as motivation (Srinivasan, 1987, 1990; Schmidt & Spreng, 1996).

Information processing. The information processing approach has its roots in psychology but focuses exclusively on memory and the cognitive limits of information processing (Bettman, 1979; Schmidt & Spreng, 1996; Srinivasan, 1990; Sternthal & Craig, 1982). Here, information search is divided into an internal search and an external search. An internal search is the process by which a buyer calls on information already stored in memory, whereas an external search is the overt search for new information (Beatty & Smith, 1987; Schmidt & Spreng, 1996). Some literature suggests that an internal search is usually performed first, followed by an external search if there is insufficient memory to make a decision (Bettman, 1979). However, other literature suggest that an internal and external may occur nearly simultaneously with an internal search occurring, followed by an external if memory is insufficient and then jumping back to internal.

There are a number of factors that influence a buyer's cognitive limits to process information. One is the buyer's ability to choose and gather relevant information, understand it, and remember it (e.g., Bettman, 1979; Petty & Cacioppo, 1986). Another factor is prior knowledge, experience and familiarity—what they have stored in memory—affects the degree to which a buyer allocates resources to performing an external search and processing new information (Srinivasan, 1990). Finally, the extent of information search depends on a combination of motivation and ability where higher ability increases information search activity (Srinivasan, 1987).

Green Consumption

Peattie (2010) conceptualizes "green consumption" broadly as "oriented toward sustainable development" and a reflection of the United Nations Environment Programme's framing of the same issue (UNEP, 2001). This means that green consumption is consumption that not only meets basic needs and enhances human well-being, but must also be more efficient, less wasteful, and take into consideration equity across generations. In this conceptualization, efficiency and waste are to be measured based on the full life cycle of material goods, while equity must include specific consideration of environmental impact and risk to human health.

Much of the research on environment-friendly consumption has sought characterize green consumers and identify the factors that influence their behavior. Here we summarize the factors that have been posited to influence behavior presented in Peattie's (2010) review of green consumption, which include those related to economics, socio-demographics, environmental perceptions, perceptions of responsibility and control, lifestyle and habits, identity and personality, consumption context, spatial dimension, social processes, social norms about the environment, and mass media.

- Financial incentives to induce environment-friendly consumption has been found to promote desired behavior (Stern, 1999; Eriksson, 2004; Bartelings & Sterner, 1999). Although costs and benefits can play an important role in environment-friendly consumption, factors such an awareness of specific costs and benefits also can influence behavior (Sorrell et al., 2000).

- A number of studies have been conducted attempting to profile green consumers and to explain behavior using *socio-demographics,* but a review by Diamantopouos et al. (2003) has found that demographic explanations are of limited value.

- A strong relationship exists between *income* and environmental *impact*, where greater affluence is positively related with greater environmental impact (WWF Cymru, 2002).

- *Environmental knowledge* is frequently cited as being an important determinant of environmental behavior, and indeed some research suggests that. However, other research has found some inconsistencies where knowledge did not translate into action. The inconsistency found in the literature needs to be further investigated, but one possible explanation is that t a distinction needs to be made between awareness and understanding (Davies et al., 2002; Pedersen & Neergaard, 2006; Southwell et al., 2012). That is, people may have high awareness of energy efficiency but may not understand their own impact on climate change in their personal lives (Anable et al., 2006).

- Generally, research shows the influence of consumers' *beliefs* and *values* on their expression of pro-environmental behavior including green consumption (Leiserowitz et al., 2006; Dietz et al., 2005; Bullock et al., 2015; Krystallis & Chryssohoidis, 2005; Nixon et al., 2009), although there has been some inconsistency between results (Pepper et al., 2009;

Barr, 2007). An understudied yet promising area of research is the role of emotion and affect in decision-making related to the environment.

- Perceived *consumer effectiveness*—consumers' belief that their action can make meaningful change on a particular issue—has been shown to significantly influence consumer behavior (De Young, 2000; Gupta & Ogden, 2009). An understudied yet promising area of research related to this is personal sense of *responsibility*. That is, how responsible does the consumer feel for correcting or mitigating environmental ills?

- *Lifestyle,* conceptualized as the host of consumption behaviors that a person exhibits, is an important factor in whether or not an individual is a green consumer throughout his or her life or at just one moment at the time of purchase (Empacher & Götz, 2004; Leiserwitz et al, 2010). Moreover, daily *habits* are another determinant of whether individuals practice pro-environmental behaviors (Warde, 2005). Furthermore, the experience of expressing a pro-environmental behavior also plays a role in whether or not an individual will continue with that behavior (Staats et al., 2004).

- Research from environmental psychological finds that people's *self-identity* has a strong influence on the nature and extent of their pro-environmental behavior (Fekadu & Kraft, 2001). Others have found that *personality* can influence pro-environmental behavior, with traits such as extroversion, agreeableness and conscientiousness being positively related (Fraj and Martinez, 2006).

- *Contextual* or situational factors have been shown to strongly influence behaviors (Peattie, 2001). For example, Dolnicar and Grün (2009) found that a majority of consumers did not maintain their pro-environmental behavior while on vacation. Other research suggests that as people move through life stages and events, their propensity to practice pro-environmental behaviors also changes. For example, moving to a new house has been found to be a life-stage opportunity to establish greener behavior patterns (Bamberg, 2006). Finally, the physical and contextual factors in which consumers' live can influence their behaviors (e.g., whether there is a local public transit system).

- Studies have found that location matters for whether or not people are able practice pro-environmental behaviors; location varies in the products available to people and in terms of the available physical infrastructure (Munksgaard et al., 2000; Tanner et al., 2004).

- Much of our consumption behavior does not simply reflect us and our circumstances, it also reflects our social relationships and obligations. We behave not just as individuals but also as members of families, households, communities, and social networks (Grønhøj, 2006; Southwell, 2013). Moreover, research shows that the behavior of others influences our reactions to and interpretations of situations that we find ourselves in (Griskevicius et al., 2006).

- *Social norms* have been found to be an important influence on green consumption (Zukin & Maguire, 2004). The concept of social norms includes what we perceive to be common practice or normal (descriptive norms) and behaviors we perceive to be morally right or that ought to be done (injunctive social norms), both of which can have a strong influence on green consumption (Jackson, 2005). For example, recycling was widely adopted because it was perceived as normal, whereas consumption reduction has not been because it has been seen as an alternative championed by a few (Barr, 2007).

- *Mass media* can play an important role in encouraging green consumption because of the dependence of the public on the media for environmental information (Haron et al., 2005) and because of media influence on environmental awareness, knowledge, opinion, and concern (Stamm et al., 2000; Southwell & Torres, 2006).

Clearly, a number of important factors influence green consumption behavior. Next, we examine information search behavior in the context of green consumption.

Green Consumption Information Search

With respect to "big" or "strategic" decisions like purchasing a home or a car, consumers often gather information from family, friends, or commercial sources. Consumers interested in green products may seek out information related to environmental performance and may consult sources such as green consumer guides or websites. McDonald et al. (2009) found that green consumers typically engaged in this type of search behavior but that ultimately behavior and sources used varied by the type of purchase. Ecological literacy, people's ability to understand environmental issues and the impact of their purchases, is also important for green consumption (Pilgrim et al., 2007).

Energy Efficiency Labeling

Theoretical Background

In recent years, some researchers have looked at eco-labeling and certification as a way to motivate green consumption behavior. Most eco-labeling initiatives target the choice phase of the purchase decision by informing consumers about ingredients, production methods, or in-use resource efficiency. Other forms of labeling are starting to emerge, such as lifespan labeling, which gives consumers extra information about the lifespan of a product (Rex & Baumann, 2007). Indeed, labels can help to address the lack of environmental literacy among consumers and the erosion of trust between producers and consumers due to "greenwashing" (Rex & Baumann, 2007). However, some research also suggests that labeling can grab attention and interest and can stimulate additional consumption, therefore negating the desired effect to the label itself (Bougherara et al., 2005; Dosi & Morretto, 2001).

Energy Efficiency Labeling and Certification

In a recent review of the trends of energy efficiency ratings in Australia, Hurst (2012) describes what is known about the role of the consumer, the influence of such labeling on behavior, and other important factors. Below is a brief adaptation of these findings.

The consumer. As noted above, with big purchases, consumers often seek professional advice and/or other information to inform their decision. With respect to real estate property information about energy efficiency, it is likely to be very difficult for inexperienced consumers to be able to make sense of the information and how it corresponds to occupant use. Indeed, consumers often have trouble making sense of the information at hand and will sometimes buy the wrong house as a result of inaccurate cost estimates or make uneconomic retrofit investments as a results of poor recommendations (Stein & Meier, 2000). Ostensibly, the purpose of an energy efficiency rating is to inform consumers of the energy costs associated with a particular home. While the ratings are intended to be house-specific, the consumer is likely to view them as relative. Indeed, as described above, it is natural during the information search process for consumers to compare product characteristics when making a decision, and houses are no different. It is through this process of comparison that the market may begin to discern the price of a home and its worth to homebuyers.

Indeed, there is some evidence that providing environmental information such as energy efficiency ratings may be of interest to homebuyers and may be a characteristic that influences purchasing decisions. In a study on willingness to pay for energy efficient homes, Mandell and Wilhelmsson (2011) found that people reported being more willing to pay a premium for homes with many environmental features. However, another study found that homebuyers were not willing to pay a premium for energy efficient features despite interest in having them (Purdie, 2009).

While it is unclear if environmental information affects final purchasing decisions, there is evidence to suggest it is of interest to the majority of buyers. According to a 2013 National Association of Home Builders report, buyers have a "wish list" of features they want in a home. Half of homebuyers wanted a new home; they wanted to spend around $203,000 and have 2,226 (median) square feet. The report's "most wanted" list shows that buyers have a strong desire for energy efficiency and storage/organization. Ninety-four percent "must have" or "want" Energy Star–rated appliances and windows and 93 percent wanted Energy Star–rated homes (Quint, 2013).

This shows that given the option, buyers will choose these features if they are available within their price range. According to Quint (2013, p. 6):

> Although the majority of home buyers are concerned about the environment in general, most are not willing to pay more for a "greener" house. In fact, 67 percent of buyers report wanting an environment friendly home or being concerned about the environment in general, but at the same time would not pay more for such a home (or even consider the impact of building the home on the environment).

The report did say that the majority of home buyers would choose a smaller home with energy efficient options. Furthermore, we are starting to see the average square footage of homes decreasing from 2,300 square feet in the 2000s to 2,100 square feet in the 2010s (Benfield, 2012). Efficiency is starting to seep into the American homebuyer mindset. Buyers are also looking to technology to have their homes perform more efficiently. According to the aforementioned National Association of Home Builders report, roughly 50 percent of homeowners already have a programmable thermostat, and a sizable minority has or wants lighting control, an energy management system or display, and smart irrigation systems. People are learning about energy efficiency, but housing inventory options with these features nonetheless are lagging behind.

Influence on behavior. Economic theories of consumption behavior suggest that people are not likely to look for and consider energy efficiency rating information until the perceived or actual benefits outweigh the costs of not doing so. Indeed, if people are aware of and understand the long-term benefits of energy efficiency in terms of reduced costs and comfort, they may seek out this type of information when purchasing a home, but people are not likely to even consider such information until they are motivated to do so. In this context, for example, consumers would need to be able to make the link between energy efficient features and energy consumption and its relevance to their lives before we would expect them to seek such information. Information processing theory suggests that before being able to consider such information and use it in the decision making process, they also must possess the ability to do so. Indeed, research suggests that energy efficiency ratings need to be clear and understandable and the relevance of energy efficiency to the consumer must also be made clear (Dewick & Miozzo, 2002; Burdock et al., 2001).

Turning to the literature on green consumption, we find that additional factors may determine whether a homebuyer will take home energy efficiency information into account, such as social processes, social norms, and situational context. Thus, both internal and external pressures come to bear on individuals when they are confronted with energy efficiency information.

Other influences. When we consider social processes, social norms, and situational context, research suggests that awareness of the benefits of energy efficient housing is increasing, as are indications that homebuyers are willing to pay for such features (Eves & Kippes, 2010; Kriese & Scholz, 2011). However, there is still lack of consensus on whether consumers are actually willing to pay for these features, and not all segments of the homebuyer population are amenable to energy efficiency for a variety of reasons, including economics. Thus, there is confusion as to what the true value of energy efficiency is to the consumer (Gill et al., 2010).

Moreover, the influence of energy efficiency ratings on consumer decision making and market value remains understudied, in part because of the lack of penetration of such systems in the market. Furthermore, a study by Gill et al. (2010) found that energy consumption in highly energy efficient housing plays a large role in the actual amount of energy saved. This result suggests that more than just energy efficiency of a home should be advertised. Homebuyers of energy efficient properties should be made aware of how to use such features to maximize their efficiency.

Best Practice Design Principles

In an effort to help people make pro-environmental decisions, the number of decision aids has been increasing, from "carbon calculators" for food purchases (Bottrill, 2007; Chatterton et al., 2009) to seasonal forecasts for farming (Meinke & Stone, 2005). These aids often offer a lot of information, drawing on resources created for professionals (Matthies et al., 2007). That very richness, Wong-Parodi and colleagues (2014) argue, runs the risk of inundating lay users with more information than they can handle. That cognitive overload may also render them vulnerable to biases induced by how information is presented, perhaps leading them to make suboptimal decisions (Kahneman, 2011).

Drawing on principles from risk communication and human-computer interaction research (Fischhoff et al., 2011; Fischhoff, 2013; Szwajcer et al., 2009; Olson & Olson, 2003), Wong-Parodi et al. (2014) offer a simple, general approach for the design of decision aids, such as those for energy efficiency. The criteria for evaluation is that aids should

> impart *knowledge* of decision-relevant facts, allow people to integrate those facts with their values well enough to form *consistent preferences,* and achieve the *active mastery* needed to make sound inferences related to practical decision problems. (Wong-Parodi et al., 2014)

Here, Wong-Parodi and colleagues work from the premise that decision making should be informed, which requires people to understand the benefits, risks, and uncertainties of their options well enough to identify choices consistent with their personal preferences (Braddock III et al., 1999). The value of providing people with good information has been observed across a wide range of domains, including environmental, financial, and health (de Best-Waldhober et al., 2009; Ver Steegh, 2002; von Winterfeldt, 2013).

Therefore, when considering how to incorporate energy efficiency rating information into an online real estate search tool, it is critical that these features be assessed for usability (understanding, forming of consistent preferences, and active mastery). That way, we can ensure that homebuyers are not only aware of energy efficiency features but are also able to understand them and to consider them critically in their decision making process.

Influence of Energy Efficiency Labeling on Purchasing Behavior

Types of Energy Efficiency Rating Systems

Home Energy Rating System (HERS) Index. Similar to the MPG labeling for automobiles or the Energy Guide for appliances, the Home Energy Rating System Index (Figure 4-1) is the current industry standard by which home energy efficiency is measured. It is also the nationally recognized system for inspecting and calculating home energy performance. A certified home energy rater assesses the home and assigns it a relative performance score. The US Department of Energy has determined that a typical home for resale scores 130 on the HERS index, while a new home scores 100. A home with a score of 70 is 30 percent more efficient than a standard new home. A home with a score of 130 is 30 percent less efficient than a standard new home. This metric label is one opportunity for homeowners to inform their decision making process when looking to purchase a new or existing home.

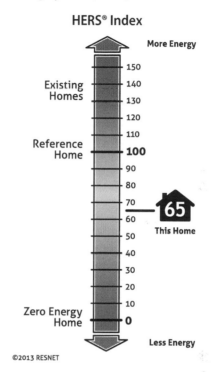

Figure 4-1. Example Home Energy Rating System (HERS) Index

©2013 RESNET

Source: Used with permission from RESNET. Available from http://www.hersindex.com.

Municipalities throughout the country have adopted a HERS index listing on real estate Multiple Listing Services (MLS) documents, which gives a homebuyers a better understanding of the particular performance of the home and adds another criteria that they can review during their purchase selection process. A growing number of states, including the following, have incorporated HERS Index Scores into MLS: Colorado, Florida, Illinois, Maine, Nebraska, New Hampshire, Oklahoma, Texas, Vermont, Virginia, and Wisconsin ("Finding Energy Efficient Homes on MLS," 2014). In addition, municipalities and states have are increasingly incorporating HERS Index Scores as a performance compliance option to building energy codes ("The HERS Index," 2015).

EPA Energy Star Home Energy Yardstick. Another option for providing energy performance information on homes comes from the EPA Energy Star Home Energy Yardstick web-based tool. This tool accounts for the number of occupants of the home as well as the type of equipment and historical utility data of the home.

Consideration should always be given based on previous energy usage of a house as compared with the projected future use. Not all homeowners operate and maintain their homes in the same manner, even identically built homes of the same age, size, construction standards, and location that are occupied at the same time. This labeling goal is intended to give a base understanding of how a particular home has performed in the past and can be a good indicator of how it may perform into the future.

Evaluation of Home Efficiency Rating Systems

To date, relatively few studies evaluating the influence of home energy rating systems on actual home buying behavior appear in peer-reviewed literature. As discussed in previous sections, the decision to buy a home is complicated and influenced by a variety of interrelated factors, so ultimately determining the impact of rating systems will require additional empirical work. Studies have found that there is an increased willingness to pay for energy efficient features, but it is not clear whether residents are actually purchasing these homes. On the one hand, research in the United Kingdom on the supply and demand for low-energy homes found that although people expressed a willingness to pay a premium for efficient features, such features were ultimately ignored when it came to negotiating the purchase price (Lovell, 2005). A recent study in Australia, however, found some promising results where homes with higher energy efficiency ratings sold for higher prices than a similar home with a lower rating over a two-year period (2005 to 2006) (Hurst, 2012).

Summary

With this chapter, we sought to present an overview of the state of knowledge on home purchasing behavior in general, and for green consumers in particular. We also presented an overview of the literature on energy efficiency rating systems and how they might be usefully designed and evaluated so that they can successfully be incorporated into real estate listings such as the MLS. Although we need more research to evaluate the role of energy efficiency rating systems in consumers' decision making and the market value of higher efficiency homes, what research has been done suggests that providing this type of information to consumers very likely could help them make informed decisions that are aligned with their values and preferences. New tools and applications to facilitate consumer information engagement, like those highlighted, offer a promising arena for investment.

Chapter References

Anable, J., Lane, B., & Kelay, T. (2006). *An evidence base review of public attitudes to climate change and transport behaviour: Final report.* Contract No. PPRO 004/006/006. London, UK: Department of Transportation.

Arndt, J. (1976). Reflections on research in consumer behavior. In B. B. Anderson (Ed.), *Advances in Consumer Research, 3,* pp. 213–21. Cincinnati, OH: Association for Consumer Research.

Bamberg, S. (2006). Is residential relocation a good opportunity to change people's travel behavior? Results from a theory driven intervention study. *Environment and Behavior, 38,* 820–40.

Barr, S. (2007). Factors influencing environmental attitudes and behaviors: A UK case study of household waste management. *Environment and Behavior, 39,* 435–73.

Bartelings, H., & Sterner, T. (1999). Household waste management in a Swedish municipality: Determinants of waste disposal, recycling and composting. *Environmental and Resource Economics, 13,* 473–91.

Bayus, B. L. (1991). The consumer durable replacement buyer. *Journal of Marketing, 55*(1), 42 –51.

Bayus, B. L., & Carlstrom, C. C. (1990). Grouping durable goods. *Applied Economics, 22*(6), 759–73.

Bazerman, M. (2001). Reflections and reviews. *Journal of Consumer Research, 27,* 4.

Beatty, S. E., & Smith, S. M. (1987). External search effort: An investigation across several product categories. *Journal of Consumer Research, 14,* 83–95.

Benfield, K. (2012). What's going on with new home sizes—is the madness finally over? [blog post]. Washington, DC: Natural Resource Defense Council. Retrieved from http://switchboard.nrdc.org/blogs/kbenfield/us_home_size_preferences_final.html

Bennett, P. D., & Harrell, G. D. (1975). The role of confidence in understanding and predicting buyers' attitudes and purchase intentions. *Journal of Consumer Research, 2,* 110–117.

Bettman, J. R. (1979). *An information processing theory of consumer choice.* Reading, MA: Addison-Wesley.

Bloch, P. H., & Richins, M. L. (1983). A theoretical model for the study of product importance perception. *Journal of Marketing, 47*(3), 69–81.

Bloom, P. N. (1989). A decision model for prioritizing and addressing consumer information problems. *Journal of Public Policy and Marketing, 8,* 161–180.

Bottrill, C. (2007). Internet-based carbon tools for behaviour change. Oxford, UK: University of Oxford Environmental Change Institute. Retrieved from http://www.eci.ox.ac.uk/research/energy/downloads/botrill07-calculators. pdf

Bougherara, D., & Grolleau, G. (2005). Can labelling policies do more harm than good? An analysis applied to environmental labelling schemes. *European Journal of Law and Economics, 19,* 5–16.

Braddock III, C. H., Edwards, K. A., Hasenberg, N. M., Laidley, T. L., & Levinson, W. (1999). Informed decision making in outpatient practice. *Journal of the American Medical Association, 282*(24), 2313–2320.

Bullock, G., Johnson, C., & Southwell, B. (2015). *Bridging the values divide: Communicating and activating diverse values to stimulate pro-environmental intentions.* Paper presented at Conference on Communication and Environment, Boulder, CO.

Burdock, E., Ritter, M., Livingston, J., & Carnes, S. (2001). Partnership for advancing technology in housing. *Forest Products Journal, 51*(3), 8.

Burnkrant, R. E. (1976). A motivational model of information processing intensity. *Journal of Consumer Research, 3,* 21–30.

Butters, G. R. (1977). Equilibrium distribution of sales and advertising prices. *Review of Economic Studies, 44,* 465–491.

Chatterton, T. J., Coulter, A., Musselwhite, C., Lyons, G., & Clegg, S. (2009). Understanding how transport choices are affected by the environment and health: Views expressed in a study on the use of carbon calculators. *Public Health, 123*(1), e45–e49.

Chaudhuri, A. & Holbrook, M. B. (2001). The chain of effects from brand trust and brand affect to brand performance: The role of brand loyalty. *Journal of Marketing, 65*(2), 81–93.

Cripps, J. D., & Meyer, R. J. (1994). Heuristics and biases in timing the replacement of durable products. *Journal of Consumer Research, 2*(2), 304–18.

Davies, J., Foxall, G. R., & Pallister, J. (2002). Beyond the intention–behaviour mythology: An integrated model of recycling. *Marketing Theory, 2,* 29–113.

de Best-Waldhober, M., Daamen, D., & Faaij, A. (2009). Informed and uninformed public opinions on CO2 capture and storage technologies in the Netherlands. *International Journal of Greenhouse Gas Control, 3*(3), 322–332.

De Young, R. (2000). Expanding and evaluating motives for environmentally responsible behavior. *Journal of Social Issues, 56,* 509–26.

Dewick, P. & Miozzo, M. (2002). Factors enabling and inhibiting sustainable technologies in construction: The case of active solar heating systems. *International Journal of Innovation Management, 6,* 257.

Diamantopoulos, A., Schlegelmilch, B. B., Sinkovics, R. R., & Bohlen, G. M. (2003). Can socio-demographics still play a role in profiling green consumers? A review of the evidence and an empirical investigation. *Journal of Business Research, 56,* 465–80.

Dietz, T., Fitzgerald, A., & Shwom, R. (2005). Environmental values. *Annual Review of Environment and Resources, 30,* 335–72.

Dolnicar, S., & Grün, B. (2009). Environmentally friendly behavior: Can heterogeneity among individuals and contexts/environments be harvested for improved sustainable management? *Environment and Behavior, 41,* 693–714.

Dosi, C. & Moretto, M. (1999). *Is ecolabeling a reliable environmental policy measure?* Fondazione Eni Enrico Mattei Working Paper No. 1999.9. Retrieved from http://ssn.com/abstract=158332

Empacher, C., & Götz, K. (2004). Lifestyle approaches as a sustainable consumption policy—a German example (pp. 190–206). In L. A. Reisch and I. Røpke (Eds.), *The ecological economics of consumption.* Cheltenham, UK: Elgar.

Eriksson, C. (2004). Can green consumerism replace environmental regulation? *Resource and Energy Economics, 26,* 281–93.

Eves, C. & Kippes, S. (2010). Public awareness of "green" and "energy efficient" residential property: An empirical survey based on data from New Zealand. *Property Management, 28*(3), 193–208.

Farley, J. U. (1964). Brand loyalty and the economics of information. *Journal of Business, 37,* 370–381.

Fekadu, Z., & Kraft, P. (2001). Self-identity in planned behavior perspective: Past behavior and its moderating effects on self-identity-intention relations. *Social Behavior and Personality, 29,* 671–86.

Finding energy efficient homes on MLS [webpage]. (2014, May 2). Retrieved from www.hersindex.com/articles/finding-energy-efficient-homes-mls/

Fischhoff, B. (2013). The sciences of science communication. *Proceedings of the National Academy of Science, 110*(Suppl. 3), 14033–14039.

Fischhoff, B., Brewer, N., & Downs, J. S. (Eds.). (2011). *Communicating risks and benefits: An evidence-based user's guide.* Washington, DC: Food and Drug Administration.

Fraj, E., & Martinez, E. (2006). Influence of personality on ecological consumer behavior. *Journal of Consumer Behavior, 5,* 167–81.

Gill, Z. M., Tierney, M. J., Pegg, I. M., & Allan, N. (2010). Measured energy and water performance of an aspiring low energy/carbon affordable housing site in the UK. *Energy and Buildings, 43,* 117–125.

Grewal, R., Mehta, R., & Kardes, F. R. (2004). The timing of repeat purchases of consumer durable goods: The role of functional bases of consumer attitudes. *Journal of Marketing Research, 41*(1), 101–15.

Griskevicius, V., Goldstein, N. J., Mortensen, C. R., Cialdini, R. B., Kenrick, D. T. (2006). Going along versus going alone: When fundamental motives facilitate strategic (non)conformity. *Journal of Personality and Social Psychology, 91,* 281–94

Grønhaug, K., Kleppe, I. A., & Haukedal, W. (1987). Observation of a strategic household purchase decision. *Psychology & Marketing, 4*(3), 239–253.

Grønhøj, A. (2006). Communication about consumption: A family process perspective on green consumer practices. *Journal of Consumer Behavior, 5,* 491–503.

Gupta, S., & Ogden, D. T. (2009). To buy or not to buy? A social dilemma perspective on green buying. *Journal of Consumer Marketing, 26,* 376–91.

Gutman, J. (1982). A means-end chain model based on consumer categorization processes. *Journal of Marketing, 46*(2), 60–72.

Hansen, F. (1972). *Consumer choice behavior: A cognitive theory*. New York: The Free Press.

Haron, S. A., Paim, L., & Yahaya, N. (2005). Towards sustainable consumption: An examination of environmental knowledge among Malaysians. *International Journal of Consumer Studies, 29*, 426–36.

Hauser, J. R., & Urban, G. L. (1986). The value priority hypothesis for consumer budget plans. *Journal of Consumer Research, 12*(4), 446–62.

Hawkins, D. I., Best, R. J., & Coney, K A. (2003). *Consumer behavior: Building marketing strategy*. Madison, WI: Irwin McGraw-Hill.

Hempel, D. J., & Punj, G. N. (1999). Linking consumer and lender perspectives in home buying: A transaction price analysis. *Journal of Consumer Affairs, 33*(2), 408–435.

Henry, P. C. (2005). Social class, market situation and consumers' metaphors of (dis)empowerment identifiers. *Journal of Consumer Research, 31*(4), 766–778.

The HERS Index as a performance path to building energy codes [blog post]. (2015). Retrieved from http://www.resnet.us/professional/main/Hers_index_and_energy_codes

Howard, A. J. (1977). *Consumer behavior: Application of theory*. New York: McGraw-Hill.

Howard, A. J., & Sheth, J. N. (1969). *The theory of buyer behavior*. New York: Wiley.

Hurst, N. (2012). Energy efficiency rating systems for housing: An Australian perspective. *International Journal of Housing Markets and Analysis, 5*(4), 361–376.

Jackson, T. (2005). Live better by consuming less? Is there a "double dividend" in sustainable consumption? *Journal of Industrial Ecology, 9*, 19–36.

Kahneman, D. (2011). *Thinking, fast and slow*. New York: Farrar Giroux & Strauss.

Kohn, M. G., & Shavell, S. (1974). The theory of search. *Journal of Economic Theory, 9*, 93–123.

Kos Koklič, M., & Vida, I. (2009). A strategic household purchase: Consumer house buying behavior. In B. Antončič (Series Ed.), *Managing Global Transitions*, 7(1), 75–96.

Kriese, U., & Scholz, R. W. (2011). The positioning of sustainability within residential property marketing. *Urban Studies, 48*(7), 1503–1527.

Krystallis A, & Chryssohoidis G. (2005). Consumers' willingness to pay for organic food: Factors that affect it and variation per organic product type. *British Food Journal, 107*, 320–343.

Leiserowitz, A. A., Kates, R. W, & Parris, T. M. (2006). Sustainability values, attitudes and behaviors: A review of multinational and global trends. *Annual Review of Environment and Resources, 31*, 413–444.

Leiserowitz, A., Maibach, E., & Roser-Renouf, C. (2010). *Americans' actions to conserve energy, reduce waste, and limit global warming: January 2010.* New Haven, CT: Yale University.

Lin, Q., & Lee, J. (2004). Consumer information search when making investment decisions. *Financial Services Review, 13*(4), 319–332.

Lovell, H. (2005). Supply and demand for low energy housing in the UK: Insights from a science and technologies approach. *Housing Studies, 20*(5), 815–829.

Mandell, S., & Wilhelmsson, M. (2011). Willingness to pay for sustainable housing. *Journal of Housing Research, 20*(1), 35–51.

Matthies, M., Giupponi, C., & Ostendorf, B. (2007). Environmental decision support systems: Current issues, methods and tools. *Environmental Modelling & Software, 22*(2), 123–127.

McDonald, S., Oates, C., Thyne, M., Alevizou, P., & McMorland, L-A. (2009). Comparing sustainable consumption patterns across product sectors. *International Journal of Consumer Studies. 33*, 137–145.

McQuiston, D. H. (1989). Novelty, complexity and importance as causal determinants in industrial buyer behavior. *Journal of Marketing, 53*(2), 66–79.

Meinke, H., & Stone, R. C. (2005). Seasonal and inter-annual climate forecasting: The new tool for increasing preparedness to climate variability and change in agricultural planning and operations. *Climatic Change, 70*(1–2), 221–253.

Mitchell, V-W. (1999), Consumer perceived risk: Conceptualizations and models. *European Journal of Marketing. 33*(1/2), 163–195.

Munksgaard, J., Pedersen, K. A., & Wier, M. (2000). Impact of household consumption on CO_2 emissions. *Energy Economics, 22,* 423–440.

Nicosia, M. F. (1966). *Consumer decision processes.* Englewood Cliffs, NJ: Prentice-Hall.

Nixon, H., Saphores J.-D. M., Ogunseitan O. A., & Shapiro A. A. (2009). Understanding preferences for recycling electronic waste in California: The influence of environmental attitudes and beliefs on willingness to pay. *Environment and Behavior, 41,* 101–124.

Olshavsky, R. W., & Wymer, W. (1995). The desire for new information from external sources. In S. Mackenzie & R. Stayman (Eds.), *Proceedings of the Society for Consumer Psychology* (pp. 17–27). Bloomington, IN: Printmaster.

Olson, G. M., & Olson, J. S. (2003), Human-computer interaction: Psychological aspects of the human use of computing. *Annual Review of Psychology, 54*(1), 491–516.

Peattie, K. (2001). Golden goose or wild goose? The hunt for the green consumer. *Business Strategy and the Environment, 10,* 187–199.

Peattie, K. (2010). Green consumption: Behavior and norms. *Annual Review of Environmental Resources, 35,* 195–228.

Pedersen, E. R., & Neergaard, P. (2006). Caveat emptor—let the buyer beware! Environmental labeling and the limitations of green consumerism. *Business Strategy and the Environment, 15,* 15–29.

Pepper, M., Jackson, T., & Uzzell, D. (2009). An examination of the values that motivate socially conscious and frugal consumer behaviours. *International Journal of Consumer Studies, 33,* 126–136.

Peter, J. P., & Olson, J. C. (2002). *Consumer behavior and marketing strategy.* New York: Irwin McGraw-Hill.

Petty, R. E., & Cacioppo, J. T. (1986). The elaboration likelihood model of persuasion. *Advances in Experimental Social Psychology. 19,* 123–205.

Pilgrim, S., Smith, D., & Pretty, J. (2007). A cross-regional assessment of the factors affecting ecoliteracy: Implications for policy and practice. *Ecological Applications, 17*(6), 1742–1751.

Punj, G. N., & Brookes, R. (2002). The influence of pre-decisional constraints on information search and consideration set formation in new automobile purchases. *International Journal of Research in Marketing, 19*(4), 383–400.

Punj, G. N., & Staelin, R. (1983). A model of consumer information search behavior for new automobiles. *Journal of Consumer Research, 9,* 366–380.

Purdie, A. J. (2009). *Market valuation of certified green homes: A case study of Colorado's built green and Energy Star programs* (Master's thesis). Montana State University, Bozeman, MT.

Quint, R. (2013). *What home buyers really want.* Washington, DC: National Association of Home Builders. Retrieved from https://www.nahb.org/~/media/Sites/NAHB/SupportingFiles/8/Wha/WhatHomeBuyersWant_20130430023250.ashx?la=en

Ratchford, B. T. (1982). Cost-benefit models for explaining consumer choice and information seeking behavior. *Management Science, 28,* 197–212.

Rex, E. & Baumann, H. (2007). Beyond ecolabels: What green marketing can learn from conventional marketing. *Journal of Cleaner Production, 15,* 567–576.

Rosenthal, L. (1997). Chain-formation in the owner-occupied housing market. *The Economic Journal, 107*(441), 478–488.

Rothschild, M. (1973). Models of market organization with imperfect information: A survey. *Journal of Political Economy, 81,* 1283–1308.

Russo, J. E. (1988). Information processing from the consumer's perspective. In E. S. Maynes (Ed.), *The frontier of research in the consumer interest* (pp. 185–218). Columbia, MO: American Council on Consumer Interests.

Russo, J. E., & Leclerc, F. (1991). Characteristics of successful product information programs. *Journal of Social Issues, 47*(1), 73–92.

Salop, S. (1976). Information and monopolistic competition. *The American Economic Review, 66,* 240–245.

Schmidt, J. B., & Spreng, R. A. (1996). A proposed model of external consumer information search. *Journal of the Academy of Marketing Science, 24,* 246–256.

Simao, A, Densham, P. J., & Haklay, M. (2009). Web-based GIS for collaborative planning and public participation: An application to the strategic planning of wind farm sites. *Journal of Environmental Management, 90*(6), 2027–2040.

Sorrell, S., Schleich, J., Scott, S., O'Malley, E., Trace, F., . . . Radgen, P. (2000). *Barriers to energy efficiency in public and private organisations: Final Report*. Brussels, Belgium: European Commission.

Southwell, B. G. (2013). *Social networks and popular understanding of science and health*. Baltimore, MD: Johns Hopkins University Press [in conjunction with RTI Press]. http://dx.doi.org/10.3768/rtipress.2013.bk.0011.1307

Southwell, B. G., & Torres, A. (2006). Connecting interpersonal and mass communication: Science news exposure, perceived ability to understand science, and conversation. *Communication Monographs, 73*(3), 334–350.

Southwell, B. G., Murphy, J. J., DeWaters, J. E., & LeBaron, P. A. (2012). *Americans' perceived and actual understanding of energy*. RTI Press Publication No. RR-0018-1208. Research Triangle Park, NC: RTI Press. http://dx.doi.org/10.3768/rtipress.2012.rr.0018.1208

Srinivasan, N. (1987). A path analytic model of external search for information for new automobiles. *Advances in Consumer Research, 14*, 319–322.

Srinivasan, N. (1990). Pre-purchase external search for information. In V. E. Zeithaml (Ed.), *Review of marketing* (pp. 153–189). Chicago: American Marketing Association.

Staats, H., Harland, P., & Wilke H. A. M. (2004). Effecting durable change: A team approach to improve environmental behavior in the household. *Environment and Behavior, 36*, 341–67.

Stamm, K. R., Clark, F., & Eblacas, P. R. (2000). Mass communication and public understanding of environmental problems: The case of global warming. *Public Understanding of Science, 9*, 219–37.

Stein, J. R. & Meier, A. (2000). Accuracy of home energy rating systems. *Energy, 25*, 339–354.

Stern, P. C. (1999). Information, incentives and pro-environmental consumer behavior. *Journal of Consumer Policy, 22*, 461–478.

Sternthal, B., & Craig, S. C. (1982). *Consumer behavior: An information processing perspective*. Englewood Cliffs, NJ: Prentice-Hall.

Stigler, G. J. (1961). The economics of information. *Journal of Political Economy, 69*, 213–255.

Stiglitz, J. E. (1979). Equilibrium in product markets with imperfect information. *The American Economic Review, 69*, 339–345.

Swan, J. E. (1969). Experimental analysis of predecision information seeking. *Journal of Marketing Research, 6,* 192–197.

Szwajcer, E. M., Hiddink, G. J., Koelen, M. A., & van Woerkum, C. M. (2009). Written nutrition communication in midwifery practice: What purpose does it serve? *Midwifery, 25*(5):509–517.

Tanner, C., Kaiser, F. G., Kast, S W. (2004). Contextual conditions of ecological consumerism: A food purchasing survey. *Environment and Behavior, 36,* 94–111.

Telser, L. G. (1973). Searching for the lowest price. *The American Economic Review, 63,* 40–49.

United Nations Environment Program (UNEP). (2001). *Consumption opportunities: Strategies for change.* Paris: UNEP.

Ver Steegh, N. (2002). Yes, no, and maybe: Informed decision making about divorce mediation in the presence of domestic violence. *William & Mary Journal of Women and the Law, 9,* 145.

von Winterfeldt, D. (2013). Bridging the gap between science and decision making. *Proceedings of the National Academy of Science, 110*(Suppl. 3):14055–14061.

Warde, A. (2005). Consumption and theories of practice. *Journal of Consumer Culture, 5,* 131–153.

Weitzman, M. L. (1979). Optimal search for the best alternative. *Econometrica, 47,* 641–654.

Wong-Parodi, G., Fischhoff, B., & Strauss, B. (2014). A method to evaluate the usability of interactive climate change impact decision aids. *Climatic Change,126*(3–4), 485–493.

Wright, P. (1975). Consumer choice strategies: Simplifying vs. optimizing. *Journal of Marketing Research, 12,* 60–67.

WWF Cymru. (2002). *The footprint of Wales: Report to the Welsh Assembly Government.* Cardiff, UK: WWF Cymru.

Zukin, S., & Maguire, J. S. (2004). Consumers and consumption. *Annual Review of Sociology, 30,* 173–97

CHAPTER 5

Energy Efficiency 101:
Improving Energy Knowledge in Neighborhoods

Dan Curry, Claudia Squire, and Gibea Besana

Introduction

Consider the perspectives of two longtime residents of Durham, North Carolina: Janice and Irene.[1]

Janice had never been excited about getting a power bill while living in her house for 36 years. Now she cannot wait for the bill to come. "My neighbors talk about my power bill, and they comment if they see I have my lights on past the time I usually turn them off. They say, 'Miss Medlyn's power bill is going to go up!' I also told my pastor that I was so convinced that this method works, that if people tried it for 2 or 3 months and they didn't save money, I would pay them! I want people to realize they do have the resources they have. They can do it by making better use of their power."

Irene was worried that attending a neighborhood workshop might be informative but boring and not much fun. Afterward, she wrote to the workshop leader that "[t]he workshop was incredibly informative and also so fun. I've been quoting what I learned about household energy use with friends and family ever since. The projects we did were easy and fun too. My favorite was cleaning the dust and dirt out from under our refrigerator. Our trainer, Dave, said that it would make our fridge so much more efficient, we'd have to lower the temperature setting. I was amazed and impressed that last night we found that our milk was frozen and we have already changed the setting. The learning was fun and interesting, but maybe the coolest part was meeting and connecting with new neighbors. My neighbors are so cool and I am excited to know them better and to see them at the next workshop."

---

[1] Much of the material in this chapter is drawn from posts to the Clean Energy Durham blog, https://cleanenergydurham.wordpress.com.

Janice and Irene are two of several thousand neighborhood residents in over a dozen communities who have experienced a different type of energy efficiency effort. In these efforts, neighborhood residents have come together to teach each other simple but powerful ways that each of them can take control of their own utility bills. With little expenditure of household funds, these residents lowered their bills by as much as 5, 10, or even 15 percent or more and are building stronger communities at the same time.

What Janice and Irene learned is that simple changes in energy use habits and a few easy do-it-yourself projects around the home can reduce energy waste and provide immediate benefits. They also experienced firsthand what can happen when neighbors come together to help each other *learn*, *do*, and *teach* others about energy use around the home.

What follows is a discussion of the challenges of changing long-standing energy use habits; how several communities engaged residents in neighbor-to-neighbor energy education using low-cost methods, achieving measurable results; and some suggestions for how this type of energy savings strategy can reach even more communities and households.

Advancing from Structural Solutions to Behavioral Changes

Much has been written and many energy efficiency programs have been designed and implemented to make our residential buildings perform better. An entire industry (home performance) has been built around providing highly trained investigators who diagnose building deficiencies and recommend cost-effective solutions. Although we have continued to focus on the structured environment, fewer resources have been dedicated to up-fitting residents with the knowledge and tools to change long-standing habits and make better energy use decisions. Habits are difficult to change, whether they are physiological, influenced by our own routines and needs, or sociological, coming from constantly changing environmental and social influences. Addressing one without understanding the impact of the other is a prescription for disappointing results.

Many theories have been used to account for household energy consumption dating back to at least the 1970s. Traditional economic approaches combined with rational choice theory hypothesize that people will make the best decisions that they can within their budget restraints. From this perspective, behavior choices can be influenced by providing information to increase knowledge and awareness and by providing more options (Frederiks et al., 2015).

More recently, a growing body of research provides evidence to suggest that people often don't act rationally and that increasing knowledge alone is not enough to create behavior change (McKenzie-Mohr, 2000; Frederiks et al., 2015; Vigen & Mazur-Stommen, 2012). For example, Geller (1981) found that people who indicated interest in enhancing the energy efficiency of their homes and participated in a workshop on energy conservation did not change their behavior despite measured change in knowledge and attitudes.

Community-based social marketing has emerged as an alternative to the rational-choice/attitude-behavior and economic self-interest models. The community-based social marketing model is based on a five-step approach that identifies barriers to the desired behavior and designs strategies to overcome those barriers (see Vigen & Mazur-Stommen, 2012).

Based on behavioral research across these models, ideas on how to better influence behavior change related to promoting energy-efficient behaviors have been identified. These include the following:

- **Identifying barriers to the desired behavior change**, as barriers to change can be both internal (e.g., lacking a necessary skill to make a home improvement) and external (lack of local access to goods such as programmable thermostats). Understanding the specific barriers experienced by individuals is a prerequisite for change (McKenzie-Mohr, 2000).

- **Setting defaults and leveraging status quo bias**, or targeting interventions at behaviors that can be done once and forgotten, can be more effective (e.g., changing a setting on a household appliance to conserve energy) (Fredericks et al., 2015).

- **Information framing and design**, as how information is conveyed can be important over and above what information is conveyed. Information that is specific, vivid, and personalized is more likely to change behavior (Stern, 1992). In addition, messages should be clear and easy to remember. Matching presented information to the attitudes, beliefs, and behaviors of the target audience is also critical. How information is framed is important (McKenzie-Mohr, 2000; Stern, 1992). Presenting information in a way that emphasizes the losses that can occur from inaction as opposed to the savings that may occur from action, are in general, more effective in promoting the targeted activity.

For example, Kahneman & Teversky (1979) found that presenting information on a water heater wrap as a way to avoid losing money was more effective than when it was presented as a way to save money.

In addition, the source of information can impact the effectiveness of a message (Stern, 1992). A study by Craig and McCann (1978) found that messages from a trusted source yielded more requests for energy conservation information and a greater actual savings than messages initiated from a less-trusted source. Conducting formative research with the target audience can be used to identify trusted sources.

- **Offering incentives**, as incentives have proven to help promote energy-efficient behaviors, particularly if the motivation to engage in the targeted behavior is low (McKenzie-Mohr, 2000).

- **Leveraging normative social influence**, which can take the form of framing the desired energy saving practices as common and socially desirable, can help shape behavior (Fredericks et al., 2015).

In light of these behavior influence models, utilities have invested in a number of initiatives, most notably the home energy reports that are now delivered by utility companies to millions of homes each month. Most utilities also have excellent do-it-yourself (DIY) tutorials available online for doing home efficiency tasks. Home energy reports provide customers with easy-to-understand analysis of their electricity bill along with comparisons with a selected sampling of similar households within a selected geographic area. Energy reductions in the 3 to 5 percent range have been achieved from home energy reports programs (Seattle City Light, 2013).

Other behavioral programs, including neighborhood competitions, gamification platforms, classroom education, and social media and online forums are being implemented in many communities. The ACEEE Field Guide to Utility-Run Behavior Programs counted 281 such programs, many with multiple iterations (Mazur-Stommen & Farley, 2013). While there are many programs and approaches, data on the cost-effectiveness of the programs is scarce, as few have reported both the cost to deliver the program and the actual energy savings that have resulted (Mazur-Stommen & Farley, 2013).

The Clean Energy Durham Model and What We Have Learned From Working in Different Communities

Janice and Irene from the introduction to this chapter are two examples of residents who attended a unique energy education program started in Durham, North Carolina. (Because we draw here from the Clean Energy Durham experience, most locations in this chapter, unless otherwise indicated, are in North Carolina.) In 2007, Clean Energy Durham began testing different methods for engaging neighborhood residents in learning how to reduce energy use in their own homes. This model focused on using the strengths inherent in neighbor-to-neighbor communications—namely, that residents would be more likely to listen to, and more readily act on, information received from their own neighbors than they would information from the city or county government or local utility company.[2]

Following several years of trials and revisions, the program was rebranded as "Pete Street, Where Neighbors Get Energy Savings," and was offered to other communities and utilities. At its core, Pete Street is designed to engage and train neighborhood volunteers who lead two types of in-home workshops for small groups of neighbors. Basic Energy Education Workshops are one-hour sessions led by community residents. At Basic Energy Education Workshops, residents learn basic information about how homes use and lose energy. The learning is reinforced by a fun and engaging energy bingo game that is both a learning tool and a great engagement opportunity as residents form teams and build new relationships.

The second type of workshop is a hands-on workshop. At a hands-on workshop, usually in a neighbor's home, a group of neighbors learn several simple DIY projects by doing them, with their own hands, while under the watchful eye of the workshop leader who has been trained on how to run these workshops. The program materials were carefully designed to ensure that each workshop is fun, that everyone can participate, and most importantly, that neighbors get to know each other a little better.

---

[2]   In order to test its hypothesis that people would be interested in neighborhood-based energy education, in 2007 Clean Energy Durham collected 182 door-to-door surveys in two middle class Durham neighborhoods. One of the survey questions was, "On a scale of 1 to 5, with 5 being the most likely, how likely would it be that your household would participate in a 2-hour energy reduction workshop on a Saturday or a weekday evening if it were a neighborhood workshop sponsored by the neighborhood association?" The mean response was 3.55. The response for sponsorship by the city or county was 2.51; by cooperative extension, 2.64; by Duke Power (the local electricity provider), 2.42. These results strongly supported a neighborhood-based approach to educating residents about saving energy.

The Pete Street program follows a learn-do-teach model. Residents *learn* about energy use and savings at neighborhood workshops, they are encouraged to *do* energy-saving projects and behaviors at home, and then they are asked to *teach* other neighbors what they have learned and done themselves. Before the end of each workshop, all participants are asked to fill out an "I Will Do/I Will Teach" form. This commitment and follow-up work afterward are important. By asking for a commitment at the very moment learning is occurring, staff intended the program to inspire the highest likelihood of action by those who have participated. To reinforce those commitments, a follow-up survey is sent to each person several weeks later as a reminder and to document energy-saving steps that they have taken as well as document who they have shared their newfound energy saving knowledge with.

When Clean Energy Durham decided to make the Pete Street program model available for other communities, it set about creating a set of user products that would assist any local community entity or utility in implementing a neighbor-to-neighbor energy education program. This included training manuals for the volunteers, workshop facilitator guides, and many document templates that could be customized by the local host organization. As of this writing, 14 communities have used the approach to varying degrees of success.

Each local entity that implemented a Pete Street program designed a local program model around its particular interests and needs. Several examples follow.

- As described by Lynn (2013), Chatham Community Development Corporation adopted the Pete Street program as a part of a larger economic development strategy called Plugging Leaks designed to keep local money from leaking out of the local community by assisting residents in lowering their utility bills, saving money, and using those funds to invest locally. In one element of this program, Chatham Community Development Corporation contracted with the county social services department to utilize some of its low-income energy assistance funds to provide energy saving education and minor repairs to assistance fund applicants. From 2010 to 2013, Plugging Leaks engaged over 400 people in energy education alone, and another 85 made modifications to their home or helped at a hands-on workshop. Households reported saving between 10 and 50 percent of their monthly utility bill and also benefitted by learning about resources that could improve their indoor air quality, how to talk to their landlord about repairs, and making better decisions about how to stay cool or keep warm.

- As a municipal electric provider, the City of Wilson used the Pete Street model to train internal city staff members who then branched out into the community facilitating workshops and training sessions. They were also able to engage a local charter school where both teachers and parents of students were invited to participate in workshops. On average, Wilson households used 10 percent less electricity after participating in a workshop (Weiss & Cheng, 2014).

- Green Opportunities, in Asheville, helps youth and adults living in poverty get and keep jobs through a workforce development model that also focuses on green construction and renovation techniques. Green Opportunities utilized the Pete Street model to deliver basic energy education to residents in the East of the Riverway neighborhood. To do so, Green Opportunities trained youth enrolled in its job training program how to facilitate energy savings workshops and provided a small stipend for each workshop that they coordinated.

- Hands On Nashville's Home Energy Savings program (in Nashville, Tennessee) engages volunteers in improving the energy efficiency, comfort, and safety of North and South Nashville homes owned and occupied by very-low-income homeowners. Hands On Nashville used the Pete Street program to bolster its existing Home Energy Savings program and provide additional energy saving information and programming to residents.

Other communities have developed partnerships with local government sustainability offices, healthy homes initiatives, weatherization programs, cooperative extension agents, affordable and transitional housing providers, economic development agencies, and veterans' and social justice organizations.

Wherever Pete Street has been used, local staff have indicated that the program materials saved the time and expense of developing materials locally and allowed for a quick launch of their program. In most cases, the neighbor-to-neighbor approach to learning and the leaders' guides that allow neighborhood volunteers to run the workshops have been key to local adoption of the approach. Often, local agencies have already been working in neighborhoods and were looking for additional ways to increase the level of fun and engagement of residents in their program. In the case of local utility providers who have adopted the Pete Street program, a primary motivation has been the need to offer an alternative response to residents who are having trouble paying their utility bills. For example, Charles Guerry, Executive VP and General Manager with Halifax Electric Membership Corporation in North Carolina, clearly stated the reason his staff wanted to engage in a partnership with Clean Energy Durham:

> In our business, the people who are buying the most electricity are usually the lowest incomes—they have poorly insulated houses, they have little energy knowledge, they have old appliances, etc. You sit across the table from people who have little income and they are struggling and they have a $400 bill and you scratch your head about what you can do for them. (personal communication with Charles Guerry, 2013)

While Pete Street provided a comprehensive set of program materials that local organizations can use to design and launch an energy education program, the adoption of the program has been slower than anticipated. This can at least partially be attributed to unfortunate timing. Clean Energy Durham began marketing Pete Street in late 2012, just as energy efficiency stimulus programs funded by the American Recovery and Reinvestment Act of 2009 were winding up across the country. A few communities, such as Chapel Hill and Carrboro, North Carolina, saw Pete Street as a way to continue providing energy efficiency services to its residents at a lower per-household cost than their stimulus-funded energy retrofit programs. But many other communities were in a downsizing mode and not considering additional programming. In North Carolina there was also a considerable reduction in state funding available to community-based organizations, which were likely to be primary hosting entities for local energy education programs.

Along the way, the communities who did adopt the Pete Street program have discovered many opportunities for partnerships (as described earlier) and some challenges that come with delivering a very hands-on program. A few of the key challenges are described below:

- **Allocating staff time to design and deliver the energy education program.** Most organizations are staff poor and everyone has multiple job duties. Adding another program is difficult and requires dedicated funding for the additional administrative time required. Local utility providers and organizations already focused on volunteer coordination were best equipped to take on this type of effort. Several communities contracted with Clean Energy Durham to fully deliver the local program.

- **Keeping track of resident interest and participation in the energy education program.** Early on, Clean Energy Durham saw the need to develop an online platform for logging resident interest and attendance at workshops, connecting residents with other neighbors who have knowledge they are willing to share, and conducting follow-up with participants. Unfortunately the funds to develop this tracking system were unavailable, so local entities need to prepare their own systems for this.

- **Finding workshop hosts.** The Pete Street concept is that a neighborhood resident will invite a group of neighbors to his or her home for a one-hour basic energy education workshop where the neighbors will learn how their homes use and lose energy, or possibly a hands-on workshop that runs about an hour and a half, where they will do with their own hands three or four simple energy saving projects and be encouraged to do those same projects at their own homes. Finding neighborhood residents willing to host these workshops has been a challenge in some communities.

To overcome this, program managers have offered basic energy education workshops in community rooms, fellowship halls, the local library, and other locations. Another strategy has been to provide free energy kits to workshop hosts as an additional incentive to hosting a workshop. A primary advantage of hosting a hands-on workshop is that the participants, under the guidance of the workshop leader, perform several energy savings projects for the host home.

- **Attempting to use so-called viral marketing has not always been successful.** In one neighborhood in Durham, North Carolina, an initial neighborhood workshop attended by 8 residents resulted in an additional 44 residents learning how to save energy around their homes as attendees at the first workshop then hosted other workshops or taught neighbors something they learned. That's the neighbor-to-neighbor concept at its best. What Clean Energy Durham has found is that this viral expansion still requires a good bit of prodding and encouragement.

Residents today are overwhelmed with information and offers from every imaginable source. Getting residents' to act in response to an energy education initiative often requires sending them multiple invitations through a variety of media. Sources could include face-to-face or text messages from neighbors, information through neighborhood newsletters and electronic mailing lists, yard signs, utility bill inserts, and door-to-door campaigns, along with mailings and other forms of communications from program managers.

- **Tracking results of energy education efforts.** To justify program delivery costs, communities need to be able to document the reductions in energy use and ideally non-energy benefits as well. This is one of the more challenging aspects of an energy education program. Unlike other energy efficiency programs where applicants fill out an application, a behavior-based effort like Pete Street collects no such information or utility records of participants. This makes it more difficult to match up people and utility use impacts.

Having local utility providers as program partners and designing the energy tracking analysis into the local strategy from the outset are important program design considerations. Where that partnership is not available, other means to track energy use must be employed. One such tool is ResiSpeak (http://www.ResiSpeak.com), which uses compiled utility data and customer online access approvals to report on energy use changes over time.

Stand-Alone or Delivered Together With Other Energy Efficiency or Community Initiatives—What Will Engage the Most Residents?

The Pete Street program has been delivered as both a stand-alone energy education program and as a component of a community or utility energy efficiency or economic development activity. From the start, a key objective of the program has been to provide owners and renters with knowledge and the tools needed to make better decisions about energy efficiency upgrades to their homes. Just as importantly, program designers believed that the energy education program would deliver real energy reductions even if program participants made no larger-scale energy upgrades to their property. That belief has been borne out based on the results reported in Warren County and Wilson, North Carolina, described in the Challenges and Opportunities section later in this chapter. There were no other significant energy efficiency incentives in place in those communities, yet participants were able to realize actual energy savings after attending an energy education workshop (Wichman & Boyle, 2012; Weiss & Cheng, 2014).

Given that the approach is effective at lowering energy use even without additional energy efficiency upgrades to the home, perhaps the best question to ask then is, Why not include resident energy education as a component of all energy efficiency programs? After all, it appears that residents can achieve an initial 5 to 10 percent energy reduction or more following attendance at a one-hour workshop with little to no out-of-pocket expense by the resident and no additional retrofit work being undertaken. Taken a step further, why not include it in all homeownership counseling, foreclosure prevention, tenant education, self-sufficiency, and financial literacy classes since every dollar saved on utility bills is a dollar that can go toward other urgent household expenses?

Energy education can be a valuable addition to most community initiatives. So why is it not happening more frequently? And how do we drastically increase the number of residents who are exposed to it so that they can start changing their energy use habits and those of their neighbors and friends?

Clean Energy Durham's experience with the Pete Street program would suggest that scaling up participation in most communities requires linking up with other community-based initiatives. In Siler City, it was a part of an economic development initiative designed to stem the flow of assets out of the community. In Asheville the connection was with a workforce development program serving unemployed youth in low-wealth neighborhoods. In another community, the plan was to link with a healthy homes initiative that was already placing outreach staff in homes.

Another opportunity niche would be to add a stronger resident energy education component to energy efficiency services provided by utilities and private contractors. Most utilities are now providing free home visits to advise residents on priority actions they could take to make their homes perform better. Many of these visits also include direct install services for a few energy upgrades, such as energy-saving lightbulbs and weatherstripping. Similarly, home performance contractors perform home energy assessments and provide owners with upgrade recommendations. During both of these in-home events, the owners/residents are usually encouraged to be present. While there may be some discussion about energy use behaviors during these interventions, most of the focus of these visits is on the physical changes being recommended to the structure. Indeed, the skill set required to provide a high-quality educational experience to a resident may be quite different than the technical skills involved in structural assessments and upgrade recommendations.

In any event, where the opportunity exists to interact with residents one-on-one in their own home or in small groups, energy educational materials that are simple and easy to understand and use should be available. This information should include clear guidance on how to present the information in a way that will encourage the residents to follow up on some of the action tips. Programs that can stack easy to use and high-quality information, small incentives for follow-through, and some type of social interaction or peer-to-peer sharing of results would likely yield the highest participation rates and the most energy savings (Mazur-Stommen & Farley, 2013).

Next Steps—What Tools and Resources Are Needed to Increase Energy Literacy for All Residents?

The Pete Street program has relied on a short list of simple to use educational tools to engage various resident groups. Included in this short list are a Basic Energy Education (BEE) Workshop Leader's Guide, a fully scripted written guide that a volunteer workshop leader can read from to facilitate one-hour basic energy education workshops. These BEE workshops have been the primary tool for engaging residents in the program. Within these hour-long workshops, residents are taught how their homes use and lose energy with a tabletop display board. The second part of the workshop engages residents in a bingo game that is designed to reinforce and enhance their understanding of home energy use. In addition to the BEE Leader's Guide there are guides for training community volunteers, manuals for how to manage a community energy education program, a detailed manual used by the trained volunteers to conduct hands-on workshops, as well as a guide for neighborhood leaders that describes how to get neighborhood residents interested in attending and learning about home energy use. All of the Pete Street materials were developed with the idea that local program managers would want to customize the tools to fit their local needs and partnerships, although providing a complete package was also an objective. In fact, a number of communities have used the materials just as they were provided, right out of the box, so to speak, to launch a program quickly and without extensive administrative cost on program design.

The initial strategy of the Pete Street program was to rely on face-to-face learning opportunities and the exchange of information between neighborhood residents in many informal ways. For that reason, and the fact that initial design work began as early as 2006, social media applications, and even a web-based platform for the program were not anticipated. Today, many utility and community-based energy efficiency programs take full advantage of online and social media applications. Some of these applications use normative social influence—framing the desired energy saving practices as common, socially desirable, and already being done by neighbors, friends, and similar social groups—to help encourage energy savings behaviors (Mazur-Stommen & Farley, 2013).

There are also a number of privately developed social mobile applications that provide incentives for reducing energy waste. JouleBug (http://joulebug. com) is one such application. It is a mobile social energy sustainability game where people can record the activities they do to save energy. Examples include using public transportation, using reusable water bottles, and buying clothes from a thrift store. For each activity, the user gets points. At the end of each period, the app informs the user how much money he or she saved. The app encourages users to compete with friends and family on Facebook and Twitter to see who can save the most money. Users can keep up with the activities of other users.

Another is Oroeco (http://www.Oroeco.com), a web-based application that rewards users for making low-carbon-impact actions in all aspects of their lives. It tracks users' climate impacts by assigning carbon values to users' purchases, home energy use, and food and transportation choices, letting users see the impact different actions and choices have on their carbon footprint. The app tallies users' carbon imprint using scientific data, and users can see how they compare with others online.

Utilities offer a variety of behavior-based programs through online sources. Most have energy saving tips and how-to video links on their customer engagement pages. Many provide normative social comparison information in their mailed utility bills, separate monthly communications (home energy reports), or customers' personal online account pages. Utilities also routinely offer social media communication options through Facebook, Twitter, Tumblr, and blogs where customers can interact with each other and company staff.

So how do we bridge the gap from Clean Energy Durham's Pete Street program, which relies on face-to-face communications between neighbors, to approaches that take full advantage of today's social media options? Would development of an online version (or e-books) of the Pete Street leader guides and training products, with more emphasis on self-directed learning, increase the reach and impact of the approach without losing the advantages inherent in the neighbor-to-neighbor approach?

We suggest that ample benefits could be derived from development of an online version of the Pete Street program. This could take the form of a series of videos covering topics such as how to lead a basic energy education workshop in your neighborhood, or hosting a hands-on workshop for your neighbors, as well as specialized how-to videos instructing how to install a programmable thermostat, applying weatherstripping to doors and windows, or how to hire a home performance contractor. A website platform that communities could use to track participation, schedule workshop sign-ups, and catalogue resident energy upgrades following participation would be very beneficial. The website could also provide opportunities for identifying and linking to program sponsors and allowing communications between participants to foster more social interactions. Sponsor links on such a site open up considerable opportunities for expansion of incentives for participation. Retailers could be approached to provide coupons for participants or even in-store locations for workshops.

Finding the Resources to Offer Effective Energy Education Programs

Before Janice and Irene and many hundreds of residents just like them can experience the empowerment of having more control of their utility bills, someone or some entity must organize the community effort. While the approach we have been discussing engages volunteer residents to become the primary "teachers," make no mistake—the task of coordinating an energy education program requires some level of paid staff time. In most of the communities implementing a Pete Street program, this has been a part-time role for one or several staff members of the utility, local government office, or community-based organization.

Anecdotal information from the Chapel Hill, Carrboro, and Siler City Pete Street programs indicates that program delivery expenses can run in the range of $100 to $300 per household participant. The higher end of the range would include the cost of providing each participant with an energy savings kit (typically running around $30 to $40 per kit). If this expenditure can produce energy reductions in the 5–10 percent range, as reported in several Pete Street communities, the value of this work certainly compares well with more expensive weatherization and retrofit projects that cost thousands of dollars per unit.

There are probably as many different funding sources for this work as there are programs doing it. A few of the sources that have been used in Pete Street communities are described below:

- **Local utility operating budgets**—Halifax EMC and the City of Wilson trained existing administrative and operational staff who were providing utility customer services to coordinate the educational program and deliver neighborhood workshops. Both entities considered approaching the North Carolina Utilities Commission with proposals for longer-term financing of this work, but concluded that more pilot studies and impact analysis were needed before that would be worthwhile.

- **Grant funding through local government partnerships**—Clean Energy Durham worked with the Durham City-County Sustainability Office to provide Pete Street workshops to residents applying for participation in energy retrofit programs funded by the US Environmental Protection Agency (EPA) and the US Department of Energy. The Towns of Chapel Hill and Carrboro also used Department of Energy funding administered through Southeast Energy Efficiency Alliance (SEEA) to implement Pete Street programs as a follow-up to their ARRA stimulus-funded energy retrofit programs. Another Pete Street program in Lee County was funded through an EPA Environmental Justice Small Grants Program allocation.

- **Community-based economic development partnerships**—Lowering utility bills benefits all households but particularly low-income households that pay proportionately higher amounts for their energy use. Recognizing this fact, organizations such as the North Carolina Community Development Initiative have been primary supporters of development of the Pete Street model. In Siler City, delivering the Pete Street program utilized funding from the NC Rural Economic Development Center's Small Town and Economic Prosperity (STEP) grant program based on the goal of keeping dollars in the local community.

- **Philanthropic support**—There are numerous foundations that support community-based initiatives, particularly ones that focus on the needs of low-income households. The neighbor-to-neighbor approach, engagement and training of community volunteers, and the financial, health and safety, and community-building aspects of the Pete Street approach are a good fit for many foundations.

- **Utility bill assistance programs**—Utilities and communities provide utility bill assistance to low-income households where failure to pay may result in utility cutoffs or evictions. The federal Low Income Home Energy Assistance Program is one such program. Utilities often have their own assistance programs as well. In Siler City, the local Pete Street program partnered with the county Department of Social Services to utilize utility bill assistance funding to offer energy education and direct install items to referred households as a more sustainable way to use the available funding.

As an alternative to governmental, utility, and foundation funding sources, private business models should also be explored. In this arrangement, contractors provide resident education and engagement opportunities to increase demand for contracted energy efficiency services. Similar in concept to contractors purchasing targeted customer lists from other businesses, suppliers, etc., home performance contractors could fund the cost of organizing and providing incentives for neighborhood energy education sessions as a strategy to garner new prospective customers who are stimulated to make further upgrades as a result of their newfound knowledge and awareness of energy use. The impact of this community economic development model could be substantial, particularly given the lower governmental funding levels presently being experienced. Further literature research and field study is needed to determine just how effective this model could be.

Challenges and Opportunities of Tracking the Outcomes of Energy Education Initiatives

As with most energy efficiency efforts, impacts of behavior-based energy education initiatives must be measured and costs justified against those impacts. Impacts include the direct reduction in energy use by participant households and a host of non-energy impacts such as improved indoor air quality, comfort, and improved longevity of equipment.

Reporting on the actual energy use reductions resulting from a behavior-based energy education program is dependent on having access to energy use data from before and after the initiative. While an energy retrofit program, hardware rebate program, or a direct install effort will have a fairly clear date at which the energy efficiency change has taken place, behavior changes may have more variability. Some residents may go home from an educational session and immediately do some simple projects or change some behaviors, whereas others may implement changes incrementally or even after attending several sessions over a span of months. So tracking behavior-based changes may require longer tracking timeframes spanning multiple seasons.

Clean Energy Durham was fortunate to have strong utility partners in two of the Pete Street community initiatives so that utility bills could be analyzed. The University of North Carolina Environmental Finance Center (UNCEFC) conducted studies of both of these programs using utility billing data from the local electricity provider. To assess the impact of energy education workshops, UNCEFC compared billing records for workshop participants with billing records of the service population that did not attend a workshop. UNCEFC applied a difference-in-difference regression methodology to estimate a treatment effect of the workshop including a variety of commonly used control variables. The result was an estimated impact of attending a workshop relative to not attending a workshop.

In the first of these two programs, Clean Energy Durham partnered with Halifax Electric Membership Corporation (HEMC) to undertake a one-year pilot program of energy savings workshops in Warren County, North Carolina. This program ran from late summer of 2011 through the spring of 2012. Although over 100 residents attended at least one energy education session, only 43 households could be matched by UNCEFC to HEMC billing data or fell within the observation period. For those 43 households, billing analysis indicated that these households reduced their electricity bills after attending any type of workshop by roughly 7.5 percent compared with non-attending households. Even more significantly, members of households who attended a hands-on workshop where the attendees actually did several do-it-yourself energy reduction projects in the home hosting the workshop reduced their electricity use in their own homes by an average of 17.5 percent compared with nonattending households (Wichman & Boyle, 2012).

In the second analyzed program, the City of Wilson used the Pete Street program materials to undertake energy education sessions for residents during 2012 and 2013. Wilson provides electricity among other municipal utility services. The UNCEFC study of this program compared the electricity usage of 228 workshop attendees to that of the general service population. The analysis found that Pete Street participants, on average, used 10 percent less electricity after participating in the program—a monthly avoided energy expenditure of roughly $18 per household (Weiss & Cheng, 2013).

These two studies provide an indication of the type of impacts that a neighbor-to-neighbor energy education program can have. More studies with larger samples are needed in order to approach regulators with proposals to utilize rate-payer funds for this type of behavior-based education program, but the initial findings are certainly encouraging.

Although energy reduction impacts will almost always be the first metric that funders and regulators will be looking at in evaluating a proposed energy efficiency program, individual consumers are often motivated more by non-energy benefits in their decision making about energy efficiency upgrades. For example, informing consumers that keeping the coils cleaned on their refrigerator can prolong the useful life of the compressor may have more meaning to them than the fact that they may be saving $3 per year in electricity.

All households can benefit from reduced utility bills, but for low-income households the impacts are enhanced. The average US household spends 4 percent of its income on home energy costs. Low-income households spend 17 percent of their annual income on energy expenditures. Any reductions in energy expenditures by low-income families can have an impact on local economic conditions, as these households are more likely to immediately spend any money saved in the local community (Mackres, 2012).

Summary

There is little doubt that behavior-based energy efficiency program offerings can result in energy and cost savings for residents. Utilities have been delivering such programs for many years. It is also clear that lower-income households that are paying a disproportionately higher percentage of their incomes for utilities would be particularly well-suited to learning how to save energy and money and would likely invest those savings in their local community.

The Pete Street program developed by Clean Energy Durham has been designed specifically to engage neighborhood residents who are not engaging in or not eligible for other utility and community-sponsored efficiency programs. Its strengths include its low- or no-cost-to-participate design, the use of trusted neighborhood volunteers for message delivery, and strong use of normative social influence: learning is delivered to groups of neighbors who can then support and encourage each other to implement energy saving schemes.

In order to reach thousands more neighbors like Janice and Irene, who are not only learning simple ways to save energy, but also are helping drive local economic reinvestment in their communities, new engagement strategies must be employed. These strategies can and should be implemented by utilities, local governments, and community-based organizations, all of which have constituencies they serve who can benefit.

There are two specific recommendations for next steps. The first is identifying an entity to develop electronic versions of products for the energy education program. Pete Street was developed prior to the social media explosion and has not explored how to use today's communications tools as an engagement strategy. There is a sense that using tools such as Facebook, Twitter, Tumblr, Instagram, and You Tube videos could assist the viral spread of an energy education program while not diminishing the neighbor-to-neighbor advantages, but this approach needs additional pilot efforts and analysis. Second, the costs and benefits of a neighbor-to-neighbor energy education program need to be analyzed further. The results of such analysis are needed to engage with regulators about how to build these approaches into utility-funded efficiency programs.

Chapter References

Craig S., & McCann, J. (1978). Assessing communication efforts of energy conservation. *Journal of Consumer Research, 3*, 82–88.

Frederiks, E. R., Stenner, K., & Hobman, E. V. (2015). Household energy use: Applying behavioral economics to understand consumer decision-making and behavior. *Renewable and Sustainable Energy Reviews, 41*, 1385–1394.

Geller, E. S. (1981). Evaluating energy conservation programs: Is verbal report enough? *Journal of Consumer Research, 8*, 331–335.

Kahneman, D., & Tversky, A. (1979). Prospect theory: an analysis of decisions under risk. *Econometrics, 47*, 263–291.

Lynn, T. (2013). *Plugging leaks: Saving $$$, saving energy. Year end report 2013*. Pittsboro, NC: Chatham Community Development Corporation.

Mackres, E. (2012). Energy efficiency and economic opportunity [blog post]. American Council for an Energy-Efficient Economy (ACEEE). Retrieved from http://aceee.org/blog/2012/09/energy-efficiency-and-economic-opport

Mazur-Stommen, S., & Farley, K. (2013). *ACEEE field guide to utility-run behavior programs*. Washington, DC: American Council for an Energy-Efficient Economy.

McKenzie-Mohr, D. (2000). Promoting sustainable behavior: An introduction to community-based social marketing. *Journal of Social Issues, 56*(3), 543–554.

Seattle City Light. (2014). *2013 impact evaluation of Home Electricity Report Program*. Seattle, WA: Seattle City Light.

Stern, P. (1992). What psychology knows about energy conservation. *American Psychologist, 47*(10), 1224–1332.

Vigen, M., & Mazur-Stommen, S. (2012). *Reaching the "high-hanging fruit" through behavior change: How community-based social marketing puts energy savings within reach*. An ACEEE White Paper. Washington, DC: American Council for an Energy-Efficient Economy.

Weiss, J., & Cheng, Y. (2014). *Residential electricity and gas customer usage and expenditure analysis, City of Wilson, NC*. Chapel Hill, NC: UNC Environmental Finance Center.

Wichman, C. J., & Boyle, C. E. (2012). *Community conversation: A preliminary evaluation of Clean Energy Durham's neighborhood energy efficiency workshops.* Chapel Hill, NC: UNC Environmental Finance Center.

PART III
CREATIVE
OPPORTUNITIES
FOR CONSUMER
ENGAGEMENT

CHAPTER 6

ENHANCING HOME ENERGY EFFICIENCY THROUGH NATURAL HAZARD RISK REDUCTION:
Linking Climate Change Mitigation and Adaptation in the Home

Christopher S. Galik, Douglas Rupert, Kendall Starkman,
Joseph Threadcraft, and Justin S. Baker

In the absence of comprehensive federal climate policy, the task of climate change mitigation and adaptation will fall to a more diffuse set of actors. There are currently over 130 million housing units in the United States (US Census Bureau, 2015). Across the full stock of US housing, estimates suggest a net-present-value-positive residential end-use energy reduction potential of approximately 28 percent by 2020 (Bouton et al., 2010). With 64.9 percent of all US housing units classified as owner-occupied (US Census Bureau, 2014), and with owner-occupied structures possessing a significantly higher per-unit energy consumption rate than rental units (US Energy Information Administration, 2012), there is substantial opportunity for home retrofits to yield significant energy savings. For example, window retrofits, upgrades, or high-performance replacement windows can generate sizable energy savings by themselves, ranging from 15 to 27 percent depending on the location and the type of improvement (Frey et al., 2012).

Opportunities likewise exist for climate change adaptation through the reduction of risk of loss due to natural disaster. Direct losses from natural disasters have increased in recent decades (Gall et al., 2011). In the near future, flood risks in particular are likely to continue to increase due to a variety of factors, including sea level rise, increased precipitation, and a growing population in at-risk urban areas (Botzen et al., 2009; Botzen & van den Bergh, 2009). Losses from these and other natural disasters can be reduced through installation or adoption of a number of structural and behavioral activities or practices (Kousky, 2012). For instance, window treatments such as shutters can provide a source of risk mitigation against wind hazards (tornadoes, hurricanes). Precise data on the benefits of these and other individual home retrofits in stemming losses from natural disasters are difficult to come by (Kreibich et al., 2005), but anecdotal information in the literature suggests that home-level structural modifications can reduce annual losses substantially— estimated by one study to be as high as 21 to 40 percent (Poussin et al., 2012; see also Botzen et al., 2009).

Importantly, the fundamental attributes of retrofit initiatives to reduce loss from climate change and weather events are similar to the attributes of increased energy efficiency retrofits. There are interesting differences in the language used to promote these improvements, however, as well as in the incentive structures provided to the targeted homeowners. These differences establish a natural experiment, one that has been replicated through the proliferation of retrofit programs in recent years. This natural experiment allows for increased understanding of homeowner response well beyond the single-program or single-objective evaluations conducted in the past. Such an increased understanding is critical as governance structures and available resources constrain implementation options, possibly forcing complex tradeoffs to be made by cities and other frontline actors (Juhola et al., 2013). Successfully leveraging the lessons learned could also facilitate so-called future proofing—those activities that reduce risk associated with a host of possible future scenarios (Thornbush et al., 2013).

Home Retrofit Programs in Research and Practice

Home retrofit programs have been employed for decades to achieve an array of public policy objectives. While energy efficiency upgrades have largely become a priority at the state level, risk mitigation efforts have been primarily dependent on local codes and standards and the delivery of programs driven by federal block grant funds. Long-term binding energy savings targets exist in 24 states, using a variety of strategies including formal energy efficiency resource standards, utility-specific tailored targets, or renewable portfolio standards (ACEEE, 2015). These statewide energy targets work through utilities to leverage resources and pass potential energy reduction incentives onto the individual ratepayer, translating directly to monetary savings for both residents, who see reduction in energy bills, and utilities, which offset costs of electricity generation. From these base achievements, local sustainability offices and programs are then able to partner with utilities and build off of the public awareness generated by utility programs.

The direct benefits of enhancing energy efficiency include reducing the demand on the power generation and distribution center, reducing the consumer's energy bill and, when viewed from a broad perspective, reducing greenhouse gas emissions. Residential energy efficiency and risk mitigation retrofit programs are generally implemented by local utilities, local governments, or nonprofit organizations (Parker & Rowlands, 2007). These small groups are more effectively able to channel efforts and funds in a place-based and targeted way and recognize the subsequent benefits and savings more directly than larger organizations.

While direct drivers of these programs may be attributed to a local implementing entity, they are rarely designed, promoted, and delivered alone. One of the greatest offerings of retrofit programs is that they bring together resources—information, expertise, contractors, financing, etc.—to create a one-stop shop for home retrofits (Fuller et al., 2010). These programs frequently create some form of partnership between local and county governments, community groups, and utilities or insurance companies (Parker & Rowlands, 2007). They also partner with experts and auditors and assemble a network of contractors that are recommended to customers as trusted retrofit implementers (Fuller et al., 2010). These programs can act as local hubs, pulling resources from a number various community assets, including residents who can provide word-of-mouth marketing and input on program direction (Southwell & Murphy, 2014).

In contrast, there are few binding high-level requirements for risk mitigation in residential homes. Local government codes and standards establish what and how to build, but comprehensive goals for the improvement of existing buildings are lacking (Berke, 1996). Program funds from federal block grants are available through programs such as the Federal Emergency Management Agency (FEMA) Pre-Disaster Mitigation Program, but this mitigation funding is allocated on a semi-competitive basis and only for defined and discrete projects (McCarthy & Keegan, 2009). Most risk mitigation home retrofits tend to be retroactive to disaster, largely because before the issue is immediately salient there is a poor understanding of risk, limited pressure from citizens, and a weak integration of mitigation into planning efforts across a multitude of sectors and concerns (Godschalk et al., 2003).

Owing to these targeting, outreach, and implementation barriers, residential retrofit programs face substantial barriers to success. Homeowners are often disinterested in retrofitting their homes, and tend to downplay the risk of damage from natural hazards and put off large expenses despite potential energy savings and increased preparedness (Godschalk et al., 2003; Fuller et al., 2010). Even for residents who may be interested, the time and effort necessary to participate in a home audit and retrofit can be enough of a deterrent to prevent participation (Fuller, 2009).

In order to spread the word that a program exists and is worth participating in, a comprehensive outreach campaign is usually necessary (Stern et al., 1986). These initiatives must appropriately target homeowners' interests and provide relevant incentives (Fuller et al., 2010), but effective marketing efforts and incentive offerings can become expensive quickly. Since many programs are funded by grants and time-constrained, there is limited ability to launch and hone cutting-edge campaigns. In addition, consistent, quality delivery of services across program offerings is also of concern, as programs generally work with a large network of contractors who must be trained and knowledgeable about program offerings and policies (Fuller et al., 2010). The investment necessary to build out and implement an effective program is both streamlined by the need to offer easy and expedient services to customers and complicated by the short-lived nature of many of these programs (see box on next page).

Owing to a substantial implementation history, a large literature on adoption of home energy efficiency practices exists.[3] For example, Stern et al. (1986) provide an informative (if now dated) review of energy conservation plan design and marketing, as well as a review of program effectiveness with an emphasis on participation rates. More recently, research has focused on retrofit program design features and the attributes of homeowners who adopt the programs (Gamtessa, 2013; Hoicka et al., 2014). The behavioral underpinnings of homeowner response have likewise featured prominently in recent research (Fuller et al., 2010; Southwell & Murphy, 2014; Wilson et al., 2014), as has the role of open-home and other community-based demonstration programs in encouraging energy efficiency retrofits (e.g., Berry et al., 2014).

Within the risk reduction arena, a shift away from centralized, infrastructure-based adaptation strategies to more homeowner-centric strategies has long been discussed (e.g., Laska, 1986). Analysis of the drivers of

---

[3] See, for example, "Driving Demand for Home Energy Improvements" (Fuller et al., 2010) for a review of energy efficiency home retrofit programs. For examples of adaptation retrofit programs, see the South Carolina Safe Home program at http://doi.sc.gov/605/SC-Safe-Home/ or My Safe Florida Home at http://www.mysafeflorida.org/mysafefloridahome.html.

Examples of Efficiency and Risk Reduction
Program Implementation

Two examples of home retrofit programs are the RePower Kitsap and My Safe Florida Home programs. RePower Kitsap is an energy efficiency home retrofit program in Washington State that aimed to divert the local utility's need to build an additional substation and additional transmission lines in the area (Kelly, 2011). In 2009, Kitsap County was awarded $5 million in grant funding by the US Department of Energy. These funds fueled a partnership between Kitsap County, the cities of Bremerton and Bainbridge Island, the Washington State University Energy Program, local utilities, a local credit union, the Washington State Department of Commerce, and a nonprofit. The marketing efforts of the program proved to be hugely successful—utilizing an online energy dashboard, bold graphics, community-inspired stories, and signage in front of every participating home (Kelly, 2011). The program reached about 1,500 homes out of the initial goal of conducting 2,000 upgrades before the grant funding expired in 2013 (Kelly, 2011; Phan, 2013).

The My Safe Florida Home Program was created in the wake of the 2004–2005 hurricane season, which damaged or destroyed approximately 20 percent of the residential housing stock in Florida (Mozumder et al., 2014). The Florida Department of Financial Services partnered with local governments, nonprofits, and "wind certification entities" (WCEs) to administer the program (Florida Department of Financial Services, 2010), which was appropriated $250 million by the State of Florida in 2006. The program was advertised to single-family homeowners through a program website, press releases, hurricane fairs, a $1 million advertising campaign, and direct marketing to WCEs (Florida State University Catastrophic Storm Risk Management Center, 2010).

Free inspections conducted across the state offered participants individualized structure ratings, mitigation improvement recommendations, and information on insurance discounts to homeowners. The results deemed 98 percent of inspected homes to be in need of new roofs—the largest and most common source of non-surge losses from hurricanes (Mozumder et al., 2014). Program implementation surpassed goals, leading to the inspection of over 400,000 homes and the financing of almost 35,000 matching grants of up to $5,000 for homeowners who opted to upgrade their homes (Florida Catastrophic Storm Risk Management Center, 2010).

Out of necessity, the program offered workforce training and certification to WCEs to establish standardized implementation methods across the state. In total, the program granted $111.2 million to residents, reached 525 low-income residents, and trained 1,600 inspectors (Mozumder et al., 2014). Due to budget constraints in Florida, the program was discontinued in 2009.

individual adaptation behavior has identified a long list of factors potentially associated with retrofit decision-making (Bubeck et al., 2012). For instance, some have found that perception of climate change and future flood risk is associated with increased incidence of household risk reduction efforts (Botzen et al., 2009; Poussin et al., 2014), while others have failed to find such a link (e.g., Bubeck et al., 2012; Kreibich, 2011). Risk perception, more broadly, has been found to be affected by both previous experience with a disaster and the type of disaster (Ho et al., 2008; see also Harvatt et al., 2011). Others, meanwhile, assess the communications strategies most likely to elicit homeowner adaptation response (Bubeck et al., 2013; Siegrist and Gutscher, 2008), the failure of individuals to adopt cost-effective practices (Kunreuther, 2006), or the policy tools best suited to maximize adoption of preferred practices (Kreibich et al., 2011). In studies that assess the potential incentives available to facilitate risk reduction behavior, the role of insurance incentives receives particular attention (Botzen et al., 2009; Kreibich et al., 2011; Peng et al., 2014).

What is missing throughout this combined research and implementation experience, however, is a comprehensive analysis of efficiency and risk reduction retrofit responses and important lessons learned for policymakers in a time of heightened need and reduced resources. As effective and efficient policies to yield desired homeowner response require insight into the most successful tools available, we need to assess the lessons provided by design and delivery of both energy efficiency and risk reduction programs. There is likewise a possibility that, apart from simple learning at the programmatic level, the retrofits themselves can be combined and delivered simultaneously, leveraging program resources to achieve climate change mitigation and adaptation objectives in the home.

Making the Case for Joint Retrofit Opportunities

Most interventions to reduce residential energy use are narrowly focused. Many efforts promoting energy efficiency or energy use reduction focus on technology or educational efforts emphasizing energy use alone rather than focusing on the interconnected aspects of homeownership and residents' daily lives (Ellsworth-Krebs et al., 2015). A variety of industries, individuals, and behaviors contribute to home improvement and repair decisions, yet these elements are oftentimes overlooked or marginalized in the quest to reduce home energy consumption. Retrofit programs for risk reduction (i.e., efforts intended to make housing more resilient to natural hazards) represent one opportunity to provide information, resources, and mechanisms simultaneously to improve homeowners' uptake of energy efficiency retrofits. Although efficiency and risk reduction programs are often performed at separate and distinct times within the life of a structure, this section of this chapter assesses the feasibility of linking energy efficiency and risk reduction assessments and retrofits during the same visit.

Retrofit Activity Synergies and Overlaps

There is a range of practices with the potential to achieve gains in home energy efficiency. Energy audits can be used to evaluate opportunities to improve the energy efficiency in homes. Audits generally consist of two separate components: a home assessment and a computer analysis (US Department of Energy, 2014a). Audits can be performed by the individual homeowner or a licensed professional, though the use of a licensed professional may have the added advantage of more advanced practices such as blower door tests, thermographic scan, utility bill review, attic and crawl space inspection, insulation evaluation, and a review of appliance and lighting efficiency (US Department of Energy, 2014b).

FEMA has determined that structural retrofits can help to protect a residence from flooding and other hazards such as earthquakes and high winds (FEMA, 2014). To protect a home from flooding, for example, these retrofits include a variety of activities that span a wide range of required investment and effort, such as elevating the lowest floor, relocating the structure to higher ground, wet or dry flood-proofing, or installation of barrier systems. Possible retrofits to protect a home from hurricane damage include protection of windows through use of impact resistant material or storm shutters, installation of hurricane-rated doors and garage doors, reinforcement of wall-to-foundation connections, and evaluation and reinforcement of roofs (Institute for Business & Home Safety, 2002).

Enhancing the training and awareness of energy efficiency auditors and risk reduction evaluators provides an opportunity to conduct a collaborative evaluation for energy efficiency and risk reduction measures. Table 6-1 provides an overview of energy efficiency and risk reduction measures performed during a typical site visit. The items in Table 6-1 are grouped by the type of evaluation (energy efficiency or risk reduction) but are not arranged in any order of either complexity or priority. We determined the minimum skill level for the individual performing each task and divided the tasks into two skill categories: certification or design professional/licensed contractor.

Based upon a review of the benefits that can be gained from combining the inspections, we recommend that policymakers and organizational leaders organize cultural and organizational change and implement consolidated inspection efforts. To ensure the success of such a program, educational outreach to the design professionals, energy auditors, finance sources, and residents of the community have to be coordinated to maximize the return on investment. As indicated in Table 6-1, retrofits for risk reduction normally

Table 6-1. Overview of potential energy efficiency and risk reduction assessment and retrofit synergies

Task	Target Objective		Skill Required to Combine Inspections	
	Energy Efficiency	Risk Reduction	Certification	Design Professional/ Licensed Contractor
Appliance evaluation	X		X	
Blower door test	X		X	
Computer analysis	X		X	
Inspect attic and crawl space	X		X	
Insulation verification	X		X	
Lighting evaluation	X		X	
Thermographic scan	X		X	
Utility bill review	X		X	
Barrier systems		X		X
Base floor elevation		X		X
Dry floodproofing		X		X
Exterior door evaluation		X	X	
Garage door evaluation		X	X	
Home demolition/ reconstruction		X		X
Home relocation		X		X
Roof evaluation		X		X
Wall to foundation evaluation		X	X	
Wet floodproofing		X		X
Window evaluation		X	X	

require the skill set of a design professional and licensed contractor for effective implementation. In the event all of the qualifications are not within the same organization, partnerships, joint ventures, and prime-sub contracts can be used to form a qualified team to consolidate the evaluations.

Retrofit Stakeholders and Markets

Apart from the simple technical viability of combining energy efficiency and risk reduction activities, there are also important data needs regarding the participants and targets of these activities. The first is better understanding the various stakeholders in the home retrofit decision making and implementation process. Homeowners are the immediate benefactor of both energy efficiency and risk reduction retrofits. But there are also important costs faced by homeowners that must be better understood and accounted for (e.g., search costs and opportunity costs). Another group of critical stakeholders includes the trades directly associated with the retrofit market, including design, installation, construction, and manufacturing industries. These are the industries and individuals responsible for physically implementing retrofits. The incentives and constraints they face, be it limited demand for a particular product or service or a shortage of time to train or be certified for a specific task or technology, will have a strong influence on the availability of a retrofit service in a given area.

These two groups—individual homeowners and retrofit service providers and trades—constitute both the demand and supply of retrofit services, and are thus arguably the most critical to implementing risk reduction and energy efficiency services in the existing stock of housing. Facilitating and impacted by the emergence of a larger market is a wider host of stakeholders.

From a risk reduction perspective, Warner et al. (2009) found that the insurance industry and the public sector could benefit from collaborating to promote risk reduction. In addition to local, state, and national government interests in risk reduction home retrofits, support also exists at the international level (Office of the Press Secretary of the White House, 2012). Efficiency retrofits are similarly relevant to electric utilities and co-ops and are championed by a wide variety of NGOs as means to achieve larger environmental objectives. Importantly, the costs and benefits of home retrofits will vary across these various stakeholder groups, creating a complex set of incentives that will ultimately define the market for retrofit activities.

Another data need pertains to the market for retrofits. Typically, the audiences for these programs are owners of single-family homes. These homeowners have the greatest incentive to invest in their property. However, as Fuller highlights, it is easy to offer loans to well-educated, motivated, and credit-worthy individuals, but these people are generally the least in need of home retrofits (Fuller, 2009). Low-income properties are less commonly included in community-wide efforts because they encounter a less financially viable set of financing requirements and often a different list of needed upgrades (Fuller, 2009). Multifamily and rental housing is also difficult to account for and often excluded from program offerings (Fuller, 2009). Even so, low participation rates in energy efficiency programs and the limited number of existing mitigation programs indicate that there is still substantial savings and mitigation potential in the residential market. Therefore, catering the approach and measure offerings to the cultural, social, and economic concerns of the target audience is critical for program success. Retrofits are a tough sell, and "significant resources and creativity need to go into promoting home energy improvements to increase participation rates" (Fuller et al., 2010).

As with stakeholders, markets for both efficiency and retrofit services will vary, presenting different barriers and opportunities for work within each. For example, as a result of continually improving building codes, newer structures tend to need less risk reduction than older ones. However, risk reduction must be a result of a concerted effort for all construction projects regardless of age and regardless of classification as single family or multifamily. It is more efficient to build risk reduction measures into the construction process than it is to retrofit structures. Furthermore, building codes vary by location to account for different conditions such as peak gust wind speeds (Figure 6-1). The existence of multiple markets for retrofits complicates the development of a single, one-size-fits-all type approach, necessitating instead careful consideration on program design and delivery.

The Importance of Communication in Program Design

Given the multiple stakeholders and markets in play, communications strategy and messaging can both be important factors in encouraging individual retrofit decision-making (Bubeck et al., 2013). Very little research has evaluated the use of communication strategies to promote home retrofits for energy efficiency or risk reduction, however. Engaging residents by meeting them where they are in a kind of "fireside chat," often called "kitchen table discussions," has been found to be successful in programs such as the Vermont Community Energy Mobilization Project and the Twin Cities One Stop Program (Fuller et al., 2010).

Figure 6-1. Peak wind gust speeds in the United States

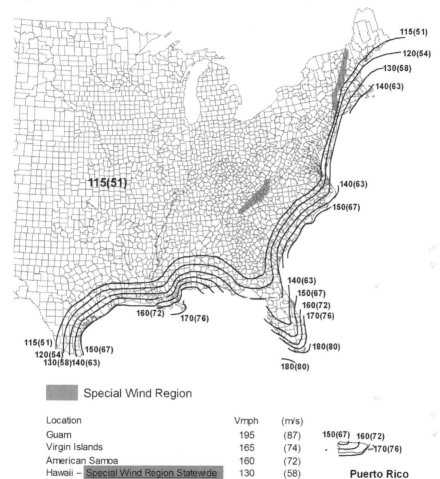

Notes:
1. Values are nominal design 3-second gust wind speeds in miles per hour (m/s) at 33 ft (10m) above ground for Exposure C category.
2. Linear interpolation between contours is permitted.
3. Islands and coastal areas outside the last contour shall use the last wind speed contour of the coastal area.
4. Mountainous terrain, gorges, ocean promontories, and special wind regions shall be examined for unusual wind conditions
5. Wind speeds correspond to approximately a 7% probability of exceedance in 50 years (Annual Exceedance Probability = 0.00143, MRI = 700 Years).

Source: American Society of Civil Engineers (2013), used with permission.

These personal meetings bring an intimate, catered feeling to the interests of the program and how they apply to potential participants. In Washington, DC, the WeatherizeDC program found notable success utilizing the strategies of political campaign organizing tools to garner participation through micro-targeting home energy improvement candidates based on demographic information (Fuller et al., 2010). More broadly, these programs commonly channel the energy generated by the communities themselves, exploiting the "snowball effect" of participation that takes place as community members engage with the program, have positive experiences and then share their experience with their friends and neighbors (Southwell & Murphy, 2014).

Other research suggests that providing information about energy efficiency recommendations often results in greater homeowner knowledge but not necessarily in behavior change (Abrahamse et al., 2005). This is especially true when efficiency behaviors are inconvenient, expensive, and require an upfront investment and when there are no easy alternatives to the energy consuming behavior (Abrahamse, 2007; Steg, 2008). Even with relatively easy activities—such as purchasing efficient appliances—cost is seen as the most important factor (Gaspar & Antunes, 2011). Individuals also are much more likely to support policies that encourage efficiency (e.g., installation of efficient appliances) than policies that restrict activities (e.g., lawn watering bans; Steg, 2008). However, efficiency messages that are tailored based on audience perceptions and values can be effective in reducing home energy use, especially when those messages are paired with goal setting and tailored feedback on home energy consumption (Abrahamse et al., 2007).

Contributions from a Broader Literature

With a few notable exceptions discussed above, much of the evidence on retrofit messaging comes from research in Europe. US homeowners may have different attitudes, beliefs, and values about home retrofit behaviors than European homeowners have. Given the lack of specific relevant research within energy efficiency literature, research in similar fields—such as home safety, environmental health, climate change attitudes, and individual energy efficiency behaviors—provides a starting point for identifying communication practices that may be effective.

Several studies have explored individuals' knowledge, attitudes, and beliefs about home safety and what communication strategies may effectively promote safe behavior. These studies have examined topics such as carbon monoxide poisoning, home furnace maintenance, and sufficient indoor heating in winter among older adults. One key finding from these home safety studies is that individuals have different motivations for adopting home safety behaviors, which need to be addressed when promoting safe practices. For example, when it comes to regular furnace maintenance or proper placement of electrical generators, younger and less experienced adults are driven less by concerns for fire, electrical, and home safety than older, more experienced users are (Rupert et al., 2013; Damon et al., 2013). Older adults' reluctance to use sufficient indoor heating in the winter is explained by different factors as well: Some are driven by financial concerns whereas others are driven by a desire to display hardiness, a distrust of government recommendations (e.g., freeze warnings), or a desire to maintain routine (Tod et al., 2012).

These studies also provide insight into trusted sources for home safety information. Homeowners rated several sources as credible and trusted outlets, including realtors, gas and electric utility companies, home insurance companies, home improvement retailers (e.g., Lowes, Home Depot), and fire departments (Damon et al., 2013; Rupert et al., 2013). Individuals also rated schools as trusted sources of home safety information, although they had mixed feelings about asking schools to focus on home safety when their primary purpose was childhood education (Rupert et al., 2013).

Multiple studies also have examined why individuals do and do not protect themselves against environmental threats—such as radon gas, water and food contamination, chemical exposures, and hurricanes—and what types of messages and information sources are most likely to encourage protective behaviors. The evidence suggests that individuals' responses to environmental risk information are driven not only by personal risk perceptions, but also by past experiences with environmental hazards, access to resources (e.g., transportation, shelter), behavior of friends and family, trust in information sources, and preferences for information delivery (Eisenman et al., 2007; Fitzpatrick-Lewis et al., 2010). For example, residents who heed evacuation orders during hurricanes are more likely to trust public officials, have greater access to transportation and alternative shelter, have weaker ties to community groups, and have an evacuation plan in place (Burnside et al., 2007; Eisenman et al., 2007).

Research also suggests that certain types of messages and sources can be more effective in promoting protective behavior. Individuals are more likely to respond to environmental threats when risk information is personalized, visual, delivered narratively, builds on existing beliefs, and addresses current (rather than past) exposures (Burger & Waishwell, 2001; Burnside et al., 2007; Connelly & Knuth, 1998; Freimuth & Van Nevel, 1993; Golding et al., 1992). Likewise, individuals are more likely to adopt recommended protective behaviors when they trust the source of the recommendation (Fitzpatrick-Lewis, 2010); public officials and media outlets are rarely rated as the most credible sources, as individuals are more likely to trust information from family and friends (Angulo et al., 1997; Burnside et al., 2007).

Climate change is an issue that elicits strong emotions for many Americans, and individuals' beliefs and attitudes on this issue have been well documented by several polls and studies (Leiserowitz et al., 2015). However, researchers also have been able to categorize individuals into more meaningful audience segments based on their climate change beliefs and attitudes, which has paved the way for more targeted and effective messaging (Leiserowtiz et al., 2015; Maibach, 2015). Specifically, individuals can be classified into one of six segments that vary in terms of their perceived importance of climate change, certainty about causes, personal risk perceptions, and perceived personal responsibility to address it (Maibach, 2015). Research also demonstrates that individuals may be more or less receptive to expert recommendations depending on whether climate change is framed as an economic, health, national security, social justice, or religious issue (Leiserowitz et al., 2015; Petrovic et al., 2014).

These findings have important implications for climate change messaging. By framing the issue in a way that resonates with individuals and builds on their current knowledge, beliefs, and values, scientists and policymakers may be able to increase support for and adoption of climate change mitigation strategies (Maibach, 2015; Roser-Renouf et al., 2014). Research has demonstrated that the terminology in and delivery of climate change messaging matters. The term global warming generally elicits more support for personal and public action than the phrase climate change, and messages with emotional appeals tend to be more persuasive (Leiserowitz et al., 2014; Wong-Parodi et al., 2011). Likewise, individuals are more willing to change their own behavior (i.e., home energy use and greenhouse gas emissions) when messages emphasize that other households will be changing, too (Staats et al., 1996).

Research and Program Implementation

Decades of research in the fields of health and risk communication emphasize the importance of identifying individuals' underlying knowledge, attitudes, and perceptions and creating messages that resonate with those attitudes and perceptions in order to influence their behavior. Specifically, audience segmentation, targeted messaging, and trusted communication sources are effective, evidence-based strategies for influencing individuals, and could be used to further promote energy efficiency and climate risk reduction retrofits among homeowners (Covello, 2003; National Cancer Institute [NCI], 2004; Walsh, 1993).

Audience segmentation is an essential component of marketing and communication, and it involves narrowing down broad populations into distinct subgroups (i.e., segments) based on similarities in demographics, knowledge, attitudes, behaviors, and other characteristics (NCI, 2004; Slater, 1995, 1996). Because segments often hold different beliefs—and demonstrate varying willingness to change—segmentation empowers us to develop messages and materials that are relevant to and resonate with each specific audience segment (NCI, 2004). Segmentation also enables us to select the different communication channels and information sources preferred by each segment. The multiple separate and overlapping markets for which retrofits may serve and the multiple, diverse stakeholders they involve underscores the potential gains generated by appropriately segmenting a retrofit program's targeted audience.

Given that each audience segment is starting with a different understanding of and attitude toward home retrofits, behavior change staff often must develop messages that resonate with a specific segment's knowledge and beliefs (Kreuter & Wray, 2003; NCI, 2004). Messages that are targeted or customized based on audience characteristics are more likely to be perceived as relevant (Kreuter & Wray, 2003), and messages perceived as personally relevant are more likely to be processed thoughtfully and completely, which more often results in lasting attitudinal and behavioral changes (Cook & Flay, 1978; Kreuter et al., 1999; Petty, 1977; Petty & Cacioppo, 1981). In practice, this means creating messages that speak to individuals' current knowledge and attitudes, correct misperceptions, demonstrate how the desired behavior aligns with their values, and acknowledge the other social, economic, and political factors involved in behavior change (Covello, 2003).

Effective communication also involves customizing not only message content, but also message delivery and sources. Trustworthy information sources are a critical element of successful communication (Covello, 2003; NCI, 2004), and sources that are perceived as more credible—whether organizations or individuals—are more likely to shift audience attitudes and perceptions (Pornpitakpan, 2004; Wilson & Sherrell, 1993). Multiple factors determine whether a source is perceived as credible (Eisend, 2006), but in general, interpersonal contacts (e.g., community leaders, family members) are rated as more trustworthy and believable than organizations and media outlets, especially if those contacts are demographically and characteristically similar to the audience segment (Atkin, 2001; NCI, 2004).

What this collective body of research suggests is that the manner in which ideas are communicated matters. Different individuals will be motivated by different concerns. Irrespective of these different underlying concerns, the choice of language can itself have a great deal to do with how individuals respond. Given the multiple stakeholders and multiple markets across which retrofit activities will take place, there is both a need and an opportunity to carefully consider the role of communication strategy in retrofit program design. On one hand, properly designing communication and outreach strategy can maximize homeowner response and make better use of limited program resources. But the use of targeted communication strategies also complicates the development of singular, one-size-fits-all type national approaches.

Incentive Mechanism and Delivery

Communicating the potential benefits of retrofit activities is a critical component of program design. Perhaps equally if not more important are the tangible benefits delivered as part of the retrofit activity itself, as well as the manner in which these benefits are delivered. A first consideration is the magnitude of the benefit generated, such as the energy savings per month or the reduced loss experienced in the event of natural disaster. A second consideration is the co-benefits that accrue, such as aesthetic improvements or decreased concern over the risk of future loss. A third consideration is the timing and manner in which benefits are received. This could include both the timing of the benefit of the retrofit itself (e.g., reduced monthly electricity or insurance bill) as well as any additional incentive that seeks to promote retrofit adoption (e.g., tax credits, low-interest financing). As with communication and outreach strategy, the presence of multiple stakeholders and markets suggests that some segments may respond more positively to some incentive design and delivery mechanisms than to others.

Retrofit programs provide a variety of services and incentives to customers. Many services offered accumulate to the individual homeowners, such as information and education, free inspections or audits, access to contractors, and accessible financing. However, the services offered by retrofit programs also accrue to the community through workforce training, relationship-building between involved parties, and improved neighborhood resilience (Fuller et al., 2010). Although retrofits themselves generate direct value through either a reduced risk of future loss or reduced electricity costs, these benefits are often insufficient to motivate homeowner adoption (Kunreuther, 2006). Strategies to further incentivize retrofits include financial lending, rebates or competitions (Fuller et al., 2010, Brown & Conover, 2009).

While energy efficiency programs generally tout the energy savings of home upgrades, risk mitigation programs can find creative ways to convey direct benefit to the customer through insurance premium discounts (Sciaudone & Lavelle, 2012). These incentives aim to reduce the cost and increase the feasibility of pursuing home retrofits for individual residents. The effect of these incentives also promotes the services themselves through social capital. As participants join the program, they share their experience with their neighbors, and such comparison can promote participation (Abrahamse et al., 2005; Southwell & Murphy, 2014).

A number of external and internal considerations can also drive program offerings. FEMA and the US Department of Housing and Urban Development (HUD) offer urban resilience and hazard mitigation grants that are awarded competitively through programs such as FEMA's Pre-Disaster Mitigation Grant Program and a HUD-conceived program now operating as Rebuild by Design (FEMA, 2015; http://www.rebuildbydesign.org). Other federal agencies such as the Department of Energy have also offered funding for community programs, most recently under the auspices of the American Recovery and Reinvestment Act. Grant dollars may also originate from state funds, which can complement federal dollars or drive the programs themselves, as seen in the example programs discussed in the following section. Other common sources of funding include larger utilities, private entities, and local nonprofits that have a local incentive (whether community-based or regulatory) to promote residential housing stock upgrades and efficiency (Fuller, 2009). In addition, some energy programs have secured funding from leasing companies, manufacturers who finance equipment, and money from state settlements (Fuller, 2009).

Improving Targeting and Uptake—Leveraging the Results of a Natural Experiment

The variety of stakeholders and markets present in any discussion of home retrofit program design and delivery, be it targeted to energy efficiency or risk reduction, requires careful consideration about the communication and incentive delivery strategy employed. Research and program experience show that communication and outreach strategy play an important role. There likewise exists a tremendous assortment of mechanisms for the delivery of incentives. Generating increased insight into how these multiple considerations can increase program uptake can help to maximize the effectiveness of scarce program resources while furthering climate change mitigation or adaption goals.

At the same time, the technical feasibility of combining retrofit activities itself creates a potential opportunity to also leverage program outreach and implementation resources to increase the climate mitigation or adaptation services delivered. Whether this is leveraging efficiency work to achieve risk reduction, or leveraging risk reduction retrofits to achieve increased energy savings, the reach of both risk reduction and energy efficiency expands. The question before us, then, is how?

The relationship between program design and retrofit behavior can be explored through the use of aggregate participation data. Statistical models can be used to predict the proportion of households that retrofit within a geographic unit as a function of relevant demographic/housing variables and program-specific attributes. Relevant program attributes could include information on program structure, whether climate mitigation or adaptation were promoted as a program benefit, implicit or explicit incentive mechanisms employed to encourage household participation, program messaging techniques, response rates, homeowner activities undertaken, and costs of the program. Understanding how these specific attributes can influence the success of a program can lead to improved design and implementation of future initiatives. Other demographic variables (ethnicity, average age in the household, education, etc.) and average housing characteristics within a target region can also help explain the variation in program participation. When demographic/housing variables are interacted with program attributes in a statistical model, this can help policymakers better understand

- why different communication strategies or program elements may be more or less effective at encouraging participation of certain segments of the population
- how program elements can be altered to improve success among targeted populations
- whether synergies between different existing programs can be combined to induce additional total benefits.

This exercise is facilitated by recent federal policy, which has helped foster a number of new initiatives that could potentially serve as a natural experiment for empirical evaluation. The American Recovery and Reinvestment Act pledged $16.8 billion to the Department of Energy's Office of Energy Efficiency and Renewable Energy to support various research, development, and outreach projects related to energy efficiency improvements or renewable energy technologies ("Energy AARA Grant Information," 2015). This included approximately $5 billion for weatherization assistance, $2.8 billion for energy efficiency and conservation block grants, and $3.1 billion for state energy programs plus an additional $454 million for the Better Buildings initiative, and $21 million for community renewable energy development. The total funding will ultimately be apportioned to different programs and projects at various spatial scales.

The proliferation of energy efficiency and risk reduction programs brings with it a broad range of program types, goals, target geographic regions, and incentive structures considered. The variation in program types offers a chance to examine specific program elements and how these relate to overall program effectiveness (which can be evaluated in terms of household participation and net energy savings following Stern et al., 1986). Empirical evidence on the relative importance of these various factors can help target outreach and implementation for future programs (including alternative strategies for communicating the benefits of program participation to homeowners). If a particular program or communication strategy is more effective at inducing participation than an alternative with a higher expected benefit to the participant, then opportunities exist to leverage the successful elements of programs with lower expected payoffs.

Lessons learned through such an exercise can certainly help to improve the design and delivery of programs. At the same time, increased information is also necessary to develop a better sense of the motivations of individual homeowners and other stakeholders. In particular, answers are needed to four major research questions about homeowners' knowledge, beliefs, and perceptions before an effective communication and marketing campaign can be designed:

1. To what extent do homeowners perceive energy efficiency and climate risk reduction as related activities?

2. Whom do homeowners see as trusted information sources on energy efficiency and climate risk reduction, respectively?

3. What motivates homeowners to adopt energy efficiency and climate risk reduction retrofits, respectively?

4. What are the barriers and facilitators to energy efficiency and climate risk reduction retrofits once homeowners are interested and motivated?

The answers to these research questions lie in additional formative research with homeowners. Formative focus groups with homeowners can address the first two research questions. Given existing research on Americans' personal beliefs and political attitudes toward climate change (Maibach, 2015), it may be advisable to segment the focus groups based on climate change profiles. This means that homeowners with similar beliefs and attitudes about climate change would be in the same groups, which should eliminate prolonged discussions about climate change politics that duplicate existing research (Leiserowitz et al., 2015; Maibach, 2015). Recalling the discussion of multiple markets above, focus groups should also represent multiple geographic areas. Conducting research in multiple areas ensures both that participants are geographically diverse and that the proposed home retrofits are relevant to them.

The final two research questions are perhaps better answered through individual interviews with homeowners who have and have not adopted energy efficiency and climate risk reduction retrofits, respectively. These interviews would explore homeowners' (a) reasons and motivations for adoption or non-adoption of retrofits; (b) reasons and motivations for adopting other home improvements; (c) factors that made or would make retrofit adoption more challenging (i.e., barriers); (d) factors that made or would make retrofit adoption easier or more appealing (i.e., facilitators); and

(e) relationships—individuals or organizations—that influenced their ultimate retrofit decisions. As with the formative focus groups, interviews should be conducted in multiple different areas in order to obtain geographic diversity and ensure homeowners had an adequate opportunity to adopt the retrofits.

All told, work to better understand the drivers of retrofit behavior can only help to further the mission of multiple federal and state agencies, foundations, and private entities. From an agency perspective, better understanding the drivers of participation can help to achieve broad policy objectives through diffused response, a necessary approach given the likely absence of any centralized policy response to our most pressing problems. More specifically, this work would help to address current shortcomings in the delivery of climate change adaptation and mitigation services. Improving retrofit uptake can help to address what is seen by some as a systematic underinvestment in risk reduction (e.g., Kunreuther, 2006), while identifying market segments most likely to respond can help to maximize program response or program benefit (Fuller et al., 2010; Peng et al., 2014). In an era of increasing public resource scarcity, such work is not only timely but critical.

Chapter References

Abrahamse, W. (2007). Energy conservation through behavioral change: *Examining the effectiveness of a tailor-made approach*. Unpublished doctoral thesis, University of Groningen, The Netherlands.

Abrahamse, W., Steg, L., Vlek, C., & Rothengatter, T. (2005). A review of intervention studies aimed at household energy conservation. *Journal of Environmental Psychology, 25*(3), 273–291.

Abrahamse, W., Steg, L., Vlek, C., & Rothengatter, T. (2007). The effect of tailored information, goal setting, and tailored feedback on household energy use, energy-related behaviors, and behavioral antecedents. *Journal of Environmental Psychology, 27*, 265–276.

American Council for an Energy Efficient Economy (ACEEE). (2015). Policy brief: State energy efficiency resource standards (EERS). Retrieved May 3, 2015, from http://aceee.org/sites/default/files/eers-04072015.pdf

American Society of Civil Engineers (ASCE). (2013). Minimum design loads for buildings and other structures, ASCE/SEI 7-10. Reston, VA: ASCE.

Angulo, F. J., Tippen, S., Sharp, D. J., Payne, B. J., Collier, C., Hill, J. E., . . . Swerdlow, D. L. (1997). A community waterborne outbreak of salmonellosis and the effectiveness of a boil water order. *American Journal of Public Health, 87*(4), 580–584.

Atkin, C. K. (2001). Theory and principles of media health campaigns. In: R. E. Rice and C.K. Atkin (Eds.), *Public communication campaigns* (pp. 49–68). Thousand Oaks, CA: Sage.

Berke, P. R. (1996). Enhancing plan quality: Evaluating the role of state planning mandates for natural hazard mitigation. *Journal of Environmental Planning and Management 39*(1), 79–96.

Berry, S., Sharp, A., Hamilton, J., & Killip, G. (2014). Inspiring low-energy retrofits: The influence of 'open home' events. *Building Research and Information, 42*, 422–433.

Botzen, W. J. W., & van den Bergh, J. C. J. M. (2009). Managing natural disaster risks in a changing climate. *Environmental Hazards: Human and Policy Dimensions, 8*, 209–225.

Botzen, W. J. W., Aerts, J. C. J. H., & van den Bergh, J. C. J. M. (2009). Willingness of homeowners to mitigate climate risk through insurance. *Ecological Economics, 68*, 2265–2277.

Bouton, S., Creyts, J., Kiely, T., Livingston, J., & Naucler, T. (Eds.) (2010). *Energy efficiency. A compelling global resource.* New York: McKinsey & Co.

Brown, M.H. & Conover, B. (2009). Recent innovations in financing for clean energy, Southwest Energy Efficiency Project. Retrieved May 3, 2015, from http://www.swenergy.org/publications/documents/Recent_Innovations_in_ Financing_for_Clean_Energy.pdf.

Bubeck, P., Botzen, W. J. W., & Aerts, J. C. J. H. (2012). A review of risk perceptions and other factors that influence flood mitigation behavior. *Risk Analysis, 32,* 1481–1495.

Bubeck, P., Botzen, W. J. W., Kreibich, H., & Aerts, J. C. J. H. (2013). Detailed insights into the influence of flood-coping appraisals on mitigation behaviour. *Global Environmental Change, 23,* 1327–1338.

Burger, J., & Waishwell, L. (2001). Are we reaching the target audience? Evaluation of a fish fact sheet. *Science of the Total Environment, 277,* 77–86.

Burnside, R., Mille, D. S., & Rivera, J. D. (2007). The impact of information and risk perception on the hurricane evacuation decision-making of Greater New Orleans residents. *Sociological Spectrum, 27,* 727–740.

Connelly, N. A., & Knuth, B. A. (1998). Evaluating risk communication: Examining target audience perceptions about four presentation formats for fish consumption health advisory information. *Risk Analysis, 18*(5), 649–659.

Cook, T., & Flay, B. (1978). The temporal persistence of experimentally induced attitude change: An evaluative review. In L. Berkowitz (Ed.), *Advances in experimental social psychology.* New York: Academic Press, 1–57.

Covello, V.T. (2003). Best practices in public health risk and crisis communication. *Journal of Health Communication, 8*(Suppl1), 5–8.

Damon, S. A., Poehlman, J. A., Rupert, D. J., & Williams, P. N. (2013). Storm-related carbon monoxide poisoning: An investigation of target audience knowledge and risk behaviors. *Social Marketing Quarterly, 19*(3), 188–199.

Eisend, M. (2006). Source credibility dimensions in marketing communication: A generalized solution. *Journal of Empirical Generalisations in Marketing Science, 10,* 2.

Eisenman, D. P., Cordasco, K. M., Asch, S., Golden, J. F., & Glik, D. (2007). Disaster planning and risk communication with vulnerable communities: Lessons from Hurricane Katrina. *American Journal of Public Health, 97*, S109–S115.

Ellsworth-Krebs, K., Reid, L., & Hunter, C. J. (2015). Home–ing in on domestic energy research: "House," "home," and the importance of ontology. *Energy Research & Social Science, 6*, 100–108.

Energy ARRA grant information [website]. (2015). Appalachian Regional Comission. Retrieved from http://www.arc.gov/funding/EnergyARRAGrantInformation.asp

Federal Emergency Management Agency (FEMA). (2014). *Homeowner's guide to retrofitting: Six ways to protect your home from flooding* (FEMA P-312, 3rd Ed). Washington, DC: Department of Homeland Security.

Federal Emergency Management Agency (FEMA). (2015). Pre-disaster mitigation grant program. Retrieved May 3, 2015, from https://www.fema.gov/pre-disaster-mitigation-grant-program

Fitzpatrick-Lewis, D., Yost, J., Ciliska, D., & Krishnaratne, S. (2010). Communication about environmental health risks: A systematic review. *Environmental Health, 9*, 67.

Florida Catastrophic Storm Risk Management Center. (2010). *Hurricane mitigation inspection system study: Final report* (DFS CS RFP 09/10-10). Tallahassee, FL: Florida State University College of Business.

Florida Department of Financial Services. (2010). *My Safe Florida Home program: Operational audit.* Report No. 2010-074. Tallahassee, FL: Florida Department of Financial Services

Freimuth, V. S., & Van Nevel, J. P. (1993). Channels and vehicles of communication: The asbestos awareness campaign. *American Journal of Industrial Medicine, 23*(1), 105–111.

Frey, P., Harris, R., Huppert, M., Spataro, K., McLennan, J., Heller, J., & Heater, M. (2012). *Saving windows, saving money: Evaluating the energy performance of window retrofit and replacement.* Washington, DC: National Trust for Historic Preservation/Preservation Green Lab. Retrieved May 3, 2015, from http://www.preservationnation.org/information-center/sustainable-communities/green-lab/saving-windows-saving-money/120919_NTHP_windows-analysis_v3lowres.pdf.

Fuller, M. C. (2009). *Enabling investments in energy efficiency: A study of energy efficiency programs that reduce first-cost barriers in the residential sector.* Berkeley, CA: California Institute for Energy and Environment.

Fuller, M. C., Kunkel, C., Zimring, M., Hoffman, I., Soroye, K. L., & Goldman, C. (2010). *Driving demand for home energy improvements: Motivating residential customers to invest in comprehensive upgrades that eliminate energy waste, avoid high utility bills, and spur the economy.* Berkeley, CA: Lawrence Berkeley National Laboratory.

Gall, M., Borden, K. A., Emrich, C. T., & Cutter, S. L. (2011). The unsustainable trend of natural hazard losses in the United States. *Sustainability, 3,* 2157–2181.

Gamtessa, S. F. (2013). An explanation of residential energy-efficiency retrofit behavior in Canada. *Energy and Buildings, 57,* 155–164.

Gaspar, R., & Antunes, D. (2011). Energy efficiency and appliance purchases in Europe: Consumer profiles and choice determinants. *Energy Policy, 39*(11), 7335–7346.

Godschalk, D. R., Brody, S., & Burby, R. (2003). Public participation in natural hazard mitigation policy formation: Challenges for comprehensive planning. *Journal of Environmental Planning and Management, 46*(5), 733–754.

Golding, D., Krimsky, S., & Plough, A. (1992). Evaluating risk communication: Narrative vs. technical presentations of information about radon. *Risk Analysis, 12*(1), 27–35.

Harvatt, J., Petts, J., & Chilvers, J. (2011). Understanding householder responses to natural hazards: Flooding and sea-level rise comparisons. *Journal of Risk Research, 14,* 63–83.

Ho, M.-C., Shaw, D., Lin, S., & Chiu, Y.-C. (2008). How do disaster characteristics influence risk perception? *Risk Analysis, 28,* 635–643.

Hoicka, C. E., Parker, P., & Andrey, J. (2014). Residential energy efficiency retrofits: How program design affects participation and outcomes. *Energy Policy, 65,* 594–607.

Institute for Business & Home Safety (IBHS). (2002). *Is your home protected from hurricane disaster? A homeowner's guide to hurricane retrofit.* Retrieved April 1, 2015, from http://www.ct.gov/cid/lib/cid /app10_hurricane.pdf

Juhola, S., Driscoll, P., de Suarez, J. M., & Suarez, P. (2013). Social strategy games in communicating trade-offs between mitigation and adaptation in cities. *Urban Climate, 4,* 102–116.

Kelly, T. (2011). *RePower Kitsap: Program gives incentives for energy efficiency.* Port Orchard, WA: WestSound Home & Garden Magazine.

Kousky, C. (2012). *Informing climate adaptation: A review of the economic costs of natural disasters, their determinants, and risk reduction options* (RFF Discussion Paper 12-28). Washington, DC: Resources for the Future.

Kreibich, H. (2011). Do perceptions of climate change influence precautionary measures? *International Journal of Climate Change Strategies and Management, 3,* 189–199.

Kreibich, H., Christenberger, S., & Schwarze, R. (2011). Economic motivation of households to undertake private precautionary measures against floods. *Natural Hazards and Earth System Sciences, 11,* 309–321.

Kreibich, H., Thieken, A. H., Petrow, T., Muller, M., & Merz, B. (2005). Flood loss reduction of private households due to building precautionary measures—lessons learned from the Elbe flood in August 2002. *Natural Hazards and Earth System Sciences, 5,* 117–126.

Kreuter, M. W., & Wray, R. J. (2003). Tailored and targeted health communication: Strategies for enhancing information relevance. *American Journal of Health Behavior, 27*(Suppl 3), S227–S232.

Kreuter, M., Bull, F., Clark, E., & Oswald, D. L. (1999). Understanding how people process health information: A comparison of tailored and untailored weight loss materials. *Health Psychology, 18*(5), 487–494.

Kunreuther, H. (2006). Disaster mitigation and insurance: Learning from Katrina. *Annals of the American Academy of Political and Social Science, 604,* 208–227.

Laska, S.B. (1986). Involving homeowners in flood mitigation. *Journal of the American Planning Association, 52,* 452–466.

Leiserowitz, A., Feinberg, G., Rosenthal, S., Smith, N., Anderson, A., Roser-Renouf, C., & Maibach, E. (2014). *What's in a name? Global warming vs. climate change.* Yale University and George Mason University. New Haven, CT: Yale Project on Climate Change Communication.

Leiserowitz, A., Maibach, E., Roser-Renouf, C., Feinberg, G., & Rosenthal, S. (2015). *Climate change in the American mind: March 2015.* Yale University and George Mason University. New Haven, CT: Yale Project on Climate Change Communication.

Maibach, E. (2015). *Communicating climate change and the case for action.* Presented at Climate Leadership Conference, Washington, DC.

McCarthy, F. X., & Keegan, N. (2009). *FEMA's pre-disaster mitigation program: Overview and issues* (CRS Report RL34537). Washington, DC: Congression Research Service.

Mozumder, P., Chowdhury, A., Vásquez, W., & Flugman, E. (2014). Household preferences for a hurricane mitigation fund in Florida. *Natural Hazards Review, 16*(3), 04014031.

National Cancer Institute. (2004). *Making health communication programs work: A planner's guide.* Bethesda, MD: National Cancer Institute.

Office of the Press Secretary of the White House (2012, June 19). G20 leaders declaration [Press release]. Retrieved April 3, 2015, from https://www.whitehouse.gov/the-press-office/2012/06/19/g20-leaders-declaration

Parker, P. & Rowlands, I. H. (2007). City partners maintain climate change action despite national cuts: Residential Energy Efficiency Programme valued at local level. *Local Environment, 12*(5), 505–517.

Peng, J., Shan, X. G., Gao, Y., Kesete, Y., Davidson, R. A., Nozick, L. K., & Kruse, J. (2014). Modeling the integrated roles of insurance and retrofit in managing natural disaster risk: A multi-stakeholder perspective. *Natural Hazards, 74,* 1043–1068.

Petrovic, N., Madrigano, J., & Zaval, L. (2014). Motivating mitigation: When health matters more than climate change. *Climate Change, 126*(1/2), 245.

Petty, R. (1977). The importance of cognitive responses in persuasion. *Advances in Consumer Research, 4,* 357–362.

Petty, R., & Cacioppo, J. (1981). *Attitudes and persuasion: Classic and contemporary approaches.* Dubuque, IA: W.C. Brown.

Phan, A. (2013, September 24). RePower Energy Retrofit Program recieves six-month extension. *Kitsap Sun.* Retrieved from http://www.kitsapsun.com/news/local-news/repower-energy-retrofit-program-receives-six

Pornpitakpan, C. (2004). The persuasiveness of source credibility: A critical review of five decades' evidence. *Journal of Applied Social Psychology, 34*(2), 243–281.

Poussin, J. K., Botzen, W. J. W., & Aerts, J. C. J. H. (2014). Factors of influence on flood damage mitigation behaviour by households. *Environmental Science and Policy, 40*, 69–77.

Poussin, J. K., Bubeck, P., Aerts, J. C. J. H. & Ward, P.J. (2012). Potential of semi-structural and non-structural adaptation strategies to reduce future flood risk: Case study for the Meuse. *Natural Hazards and Earth System Sciences, 12*, 3455–3471.

Roser-Renouf, C., Stenhouse, N., Rolfe-Redding, J., Maibach, E., & Leiserowitz, A. (2014). Engaging diverse audiences with climate change: Message strategies for global warming's six Americas. In: A. Hansen & R. Cox (Eds.), *The Routledge handbook of environment and communication.* New York: Routledge, 368–400.

Rupert, D. J., Poehlman, J. A., Damon, S. A., & Williams, P. N. (2013). Risk and protective behaviours for residential carbon monoxide poisoning. *Injury Prevention, 19*(2), 119–123.

Sciaudone, J., & Lavelle, F. (2012). Evolution of insurance incentives for wind-resistant construction since Hurricane Andrew. *Advances in Hurricane Engineering:* 200–211. http://dx.doi.org/10.1061/9780784412626.019

Siegrist, M., & Gutscher, H. (2008). Natural hazards and motivation for mitigation behavior: People cannot predict the affect evoked by a severe flood. *Risk Analysis, 28*, 771–778.

Slater, M. D. (1995). Choosing audience segmentation strategies and methods for health communication. In: E. Maibach, & R. Parrott (Eds.), *Designing health messages: Approaches from communication theory and public health practice.* Thousand Oaks, CA: Sage Publications, 186–198.

Slater, M. D. (1996). Theory and method in health audience segmentation. *Journal of Health Communication, 1*(3), 267–283.

Southwell, B. G., & Murphy, J. (2014). Weatherization behavior and social context: The influences of factual knowledge and social interaction. *Energy Research & Social Science, 2*, 59–65.

Staats, H. J., Wit, A. P., & Midden, C. Y. H. (1996). Communicating the greenhouse effect to the public: Evaluation of a mass media campaign from a social dilemma perspective. *Journal of Environmental Management, 46*(2), 189–203.

Steg, L. (2008). Promoting household energy conservation. *Energy Policy, 36,* 4449–4453.

Stern, P. C., Aronson, E., Darley, J. M., Hill, D. H., Hirst, E., Kempton, W., & Wilbanks, T. J. (1986). The effectiveness of incentives for residential energy conservation. *Evaluation Review, 10,* 147–176.

Thornbush, M., Golubchikov, O., & Bouzarovski, S. (2013). Sustainable cities targeted by combined mitigation–adaptation efforts for future-proofing. *Sustainable Cities and Society, 9,* 1–9.

Tod, A. M., Lusambili, A., Homer, C., Abbott, J., Cooke, J. M., Stocks, A. J., & McDaid, K. A. (2012). Understanding factors influencing vulnerable older people keeping warm and well in winter: A qualitative study using social marketing techniques. *BMJ Open, 2*(4), e000922.

US Census Bureau. (2015). National summary report and tables: 2013 [Data file]. Recovery Act state memos: North Carolina. Retrieved June 1, 2010, from US Department of Energy: http://energy.gov/sites/prod/files/edg/recovery/documents/Recovery_Act_Memo_North_Carolina.pdf

US Department of Energy. (2014a).#AskEnergysaver: Home energy audits. Retrieved April 1, 2015, from http://energy.gov/articles/askenergysaver-home-energy-audits

US Department of Energy. (2014b). Professional home energy audits. Retrieved April 1, 2015, from http://energy.gov/energysaver/articles/professional-home-energy-audits

US Energy Information Administration. (2012). Summary household site consumption and expenditures in the US – totals and intensities, 2009 (Table CE1.1). Retrieved January 7, 2015, from http://www.eia.gov/consumption/residential/data/2009/c&e/summary/xls/CE1.1 Summary Site.xlsx

Walsh, D.C., Rudd, R.E., Moeykens, B.A., & Moloney, T.W. (1993). Social marketing for public health. *Health Affairs, 12*(2), 104–119.

Warner, K., Ranger, N., Surminski, S., Arnold, M., Linnnerooth-Bayer, J., Michel-Kerjan, E., . . . Herweijer, C. (2009). *Adaptation to climate change: Linking disaster risk reduction and insurance.* Geneva, Switzerland: United Nations International Strategy for Disaster Reduction Secretariat.

Wilson, C., Crane, L., & Chryssochoidis, G. (2014). *Why do people decide to renovate their homes to improve energy efficiency?* Working Paper 160. Norwich, UK: Tyndall Centre for Climate Change Research, University of East Anglia.

Wilson, E. J., & Sherrell, D. L. (1993). Source effects in communication and persuasion research: A meta-analysis of effect size. *Journal of the Academy of Marketing Science, 21*(2), 101–112.

Wong-Parodi, G., Dowlatabadi, H., McDaniels, T., & Ray, I. (2011). Influencing attitudes toward carbon capture and sequestration: A social marketing approach. *Environmental Science & Technology, 45*(16), 6743–6751.

CHAPTER 7

Leveraging the Employer-Employee Relationship to Reduce Greenhouse Gas Emissions at the Residential Level

Charles Adair, Jennifer Weiss, and Jason Elliott

Introduction

Under the United Nations Framework Convention on Climate Change, 196 countries have committed to a common goal of stabilizing greenhouse gas concentrations in the atmosphere to avoid significant human disruption of the earth's climate system (UNFCCC, 2015). To meet this goal, according to the 2014 Intergovernmental Panel on Climate Change report, greenhouse gas emissions must be reduced by 40 to 70 percent against a 2010 baseline in order to keep warming under a safe threshold of 2 degrees Celsius (IPCC, 2014). The US Environmental Protection Agency (EPA) estimates approximately 17 percent of national greenhouse gas emissions come from electricity and fuel use in the residential sector, making it an important area for reductions (US EPA, 2015a).

While it is possible to decrease greenhouse gas emissions from use of electricity by shifting toward clean energy sources, our ability to reach zero emissions is limited by cost and available technology. Reducing emissions quickly and significantly therefore requires action both on the production side (burning fewer carbon-intensive fuels) and on the consumption side (reducing energy use). Within the residential sector, this encompasses both actions that reduce energy use overall (energy conservation and energy efficiency) and actions that reduce demand for energy from fossil fuels by displacing it with renewable energy (e.g., solar). In this chapter, the term *energy use reduction* is used to describe both types of actions, which reduce demand for energy from fossil fuels on the grid.

This chapter begins by exploring barriers that keep individuals from taking action to reduce their energy use at home, shares strategies for overcoming these barriers, and identifies groups that are currently working to remove these barriers at the residential level. Then, using evidence from pilot programs and other research at Duke University, this chapter identifies employers as a group that could play a significant role in reducing residential energy use, provides recommendations for creating employer-based energy use reduction programs, and shares lessons learned. This chapter concludes with an analysis of the advantages and challenges of implementing an employer-based energy use reduction program and identifies areas that require further study.

Energy Use at the Residential Level

Today, there are over 130 million houses in the United States (US Census Bureau, 2015). Each uses an average of more than 900 kWh of electricity per month (US EIA, 2015). The sheer number of individual residences combined with the diversity of the building stock presents a significant challenge to reducing energy use and associated greenhouse gas emissions.

Despite these obstacles to improving the energy performance of such a large and diffuse group of buildings, recent technological advances are empowering individuals to take action. The cost of energy generated using solar photovoltaic (PV) energy has dropped more than 65 percent over the past 17 years (Feldman et al., 2014, pg. 8). LEDs use 75 percent less energy and last 25 times longer than equivalent incandescent bulbs (US DOE, 2012a). Energy Star appliances perform daily household chores at a fraction of the energy use of older appliances (US EPA, 2015b). Smart thermostats, lightbulbs, power strips, and appliances can be controlled remotely from a smartphone, enabling homeowners to track and control energy use like never before.

This technology landscape has made it possible to address both the production and consumption sides of residential energy use at the household level through a combination of energy conservation, energy efficiency, and distributed renewable electricity (e.g., rooftop solar). Furthermore, energy use reduction programs have the potential to provide significant educational, economic, social, and environmental co-benefits to residents and their surrounding communities such as cost savings, comfort, and improved indoor air quality.

Seeing this opportunity, the Duke Carbon Offsets Initiative (DCOI), a program within Sustainable Duke at Duke University, set out to develop a suite of pilot programs to encourage employees to implement energy conservation, energy efficiency, and residential solar in a way that would benefit Duke University and the local community. The DCOI is tasked with helping the

University reach climate neutrality by 2024. To help achieve this goal, the DCOI develops local projects that reduce greenhouse gas emissions and also provide significant educational, economic, social, and environmental co-benefits to the university and surrounding community. Such energy use reduction programs directly align with the DCOI's mission.

Recognizing that the findings from these pilot programs could help inform program development at other schools and other employers, the Duke Endowment provided funding for an energy efficiency employee benefit pilot program and the Bass Connections in Energy program provided funding for a residential solar pilot. To develop the most effective pilot programs, the DCOI explored the barriers to implementation that prevent residents from reducing energy use within their homes and researched similar programs across the country. To date, the DCOI pilot programs have served more than 100 employees and have helped identify an initial set of best practices for employer-based energy use reduction programs.

The Energy Use Reduction Pyramid and Barriers to Implementation

Taken together, the three types of energy use reduction programs can be thought of as an energy use reduction pyramid (Figure 7-1). At the base of the pyramid, energy conservation through behavioral change (e.g., reducing energy use by turning off the

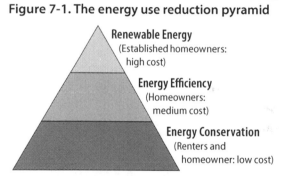

Figure 7-1. The energy use reduction pyramid

Renewable Energy
(Established homeowners: high cost)

Energy Efficiency
(Homeowners: medium cost)

Energy Conservation
(Renters and homeowner: low cost)

lights) is accessible to the largest group of people due to the low monetary cost of implementation and the ability of renters to participate (Gillingham et al., 2009). The second tier, energy efficiency (e.g., air sealing, duct sealing, insulation), costs more up front but can be implemented to some degree by most homeowners. Finally, at the top of the pyramid, renewable energy has the highest total cost and is only appropriate for homes with certain characteristics (e.g., south-facing roof, no shading for solar). In addition, reducing energy use through energy conservation and energy efficiency prior to installing renewable energy reduces the overall energy use within the home and allows the homeowner to purchase a smaller system, thereby reducing overall costs. In this chapter, we use residential solar to represent the renewable energy tier. In an ideal world, each household would be able to easily implement all energy use reduction activities available at each tier of the pyramid. In reality, it is difficult to mobilize both renters and homeowners to take action due to a number of well-established barriers to implementation. Table 7-1 summarizes common barriers (Gillingham et al., 2009, pp. 602-610).

Table 7-1. Barriers to energy conservation, energy efficiency, and renewable energy (residential solar)

Barrier	Energy Conservation	Energy Efficiency	Residential Solar
Lack of information (not knowing what to do or where to start)	X	X	X
Lack of trust of contractors	N/A	X	X
Upfront cost	N/A	X	X
Lack of access to low-interest loans	N/A	X	X
Lack of time	X	X	X
Convenience	X	X	X

N/A = not applicable

Lack of Information

Lack of information—or asymmetric information between individuals and energy service providers—is an important barrier for all three energy use reduction options. The opportunities at each tier of the pyramid are specific to each home, and many homeowners and renters do not know what to do or where to start (Gillingham et al., 2009). The many different types of potential energy conservation and efficiency actions that homeowners and renters can take within their residences makes this decision especially challenging (Frederiks et al., 2015). Solar has a more standardized installation process than energy efficiency retrofits, but many homeowners still have trouble deciding whether they should install solar and, if so, which contractor to call (Rai & McAndrews, 2012).

Lack of Trust

Furthermore, many individuals lack trust in contractors, whose expertise is frequently needed to complete energy efficiency retrofits or install solar panels. Individuals may have had poor experiences with contractors in the past, or have difficulty trusting unfamiliar companies performing costly work they know little about (Gillingham et al., 2009). Because energy efficiency and solar are typically optional home improvements, if homeowners are unsure of the quality of a contractor, they may choose to forego an installation rather than risk a poor experience (Zeelenberg & Pieters, 2002). Relative to energy efficiency retrofits, this barrier may become somewhat less important in the context of solar installations as the process becomes streamlined, but it is likely to remain a notable concern because of high costs and large number of contractors to choose from (Haynes, 2009).

Upfront Cost

The upfront costs of energy efficiency retrofits and solar are significant and prevent many homeowners from taking action, especially if low interest rate loans are not available (Borgeson et al., 2012). While some banks do offer low interest loans for both energy efficiency retrofits and solar, low interest financing is more commonly available for solar installations because many solar companies either provide these products directly or offer financing through partnerships with banks and credit unions (DOE, 2010a, pg. 3). A wide range of government incentives exist at the national, state, and local levels to reduce the upfront cost of clean energy investments (e.g., tax credits, rebates, and low interest financing policies), but these programs are frequently temporary and are not always accessible to all individuals (e.g., many programs are limited in geographic scope, and applying tax credits requires tax liability) (DSIRE, 2015). For example, North Carolina currently offers a 35 percent tax credit incentive for new solar installations, which can be taken in addition to the federal 30 percent tax credit. However, the North Carolina tax credit is set to expire in December 2015, causing uncertainty in the solar investment market (NC General Assembly, 2009, pg. 1). In some states (e.g., California), third-party financing models that have no upfront cost have enabled rapid growth in residential solar installations, providing further evidence of the importance of upfront cost as a barrier where similar no-money-down financing is not available (Corfee et al., 2014, pp. 8-12).

Lack of Time and Convenience

Finally, all three tiers require a significant time commitment on the part of the resident or homeowner. Time and convenience constraints can prevent individuals from taking action (Sorrell et al., 2010, pg. 169). Both energy efficiency retrofits and solar installations require multiple home visits from the contractor, which may necessitate time away from work for residents to meet with contractors at their home. If the contractor is trusted, however, homeowners may not feel the need to be present for the installation.

Taken together, these barriers create significant obstacles to reducing residential energy use and greenhouse gas emissions. The following sections explore strategies for reducing or eliminating these barriers, review the types of organizations that are currently working to facilitate home energy use reduction, and explore the potential for employers to incentivize home energy use reduction through employee benefit programs.

Removing Barriers to Implementation

Residential energy use reduction programs typically aim to address one or more of the barriers to implementation identified in the previous section. While many programs focus on mitigating a single barrier (e.g., tax credits reduce the financial barrier to investment), programs that remove multiple barriers simultaneously may reach a broader population (Sovacool, 2009).

Information

Providing easily accessible trusted information about the need and opportunities for home energy use reduction can mitigate the information barrier and move individuals to action. Information can be shared through access to online resources, in-person consultations, presentations from experts, community discussion, and other forums (ACEEE, 2013). For example, the nonprofit organization Clean Energy Durham employed a model in which the organization provided educational information to neighborhood champions, who then taught others in their neighborhood about energy conservation and efficiency (Clean Energy Durham, 2014). In the Clinton Climate Initiative's Home Energy Affordability Loan Program (CCI-HEAL) model, program representatives walk each participant through his or her audit report to explain the results and answer questions (Clinton Climate Initiative, 2015). Taking a different approach, the North Carolina Cooperative Extension eConservation Program provides many resources for the public regarding energy conservation and efficiency, including training videos and fact sheets (North Carolina Cooperative Extension, 2015).

Lack of Trust

Energy use reduction programs address the barrier of lack of trust in contractors by vetting contractors and other vendors to help residents identify the best local contractors. A transparent and thorough vetting process may be enough to overcome this barrier, but having a trusted program manager or program implementer could further increase the ability of the program to remove the lack of trust barrier completely. For example, a program managed by the Oregon nonprofit Clean Energy Works, now Enhabit (https://enhabit. org) requires all audit contractors to undergo a vetting process and meet specific standards to participate in their program. In the Duke Carbon Offsets Initiative's (DCOI) version of the HEAL program, pilot participants reported feeling more confident about the decision to undertake energy efficiency measures after reviewing their energy audit information with the DCOI program manager (a Duke University staff member), even knowing that all participating contractors were previously vetted.

Upfront Costs

Energy use reduction programs can reduce the upfront costs of energy efficiency or residential solar in a number of ways, including rebates, tax credits, subsidies, and group purchasing discounts. In addition, programs can set up low interest loan products or on-bill financing to help participants manage any costs they do incur (Freehling, 2011). Examples include the City of Durham's American Recovery and Reinvestment Act (ARRA)-funded weatherization program (direct subsidy), Solarize North Carolina group purchasing programs (http://solarize-nc.org/), and the South Carolina co-ops' Help My House on-bill financing energy efficiency program (Environmental and Energy Study Institute, 2013).

Lack of Time and Convenience

Finally, programs can offer general assistance to help streamline installation processes, thereby decreasing the time required of program participants and making the entire process of a home energy retrofit more convenient. For example, the CCI-HEAL program helps participants set up and schedule energy efficiency audits (Clinton Climate Initiative, 2015). Similarly, the

Oregon Enhabit program provides a web portal that residents can use to track their audit reports, bids, and other information.

Entities That Commonly Administer Energy Use Reduction Programs

Utilities, nonprofit organizations, and state and local governments are the most common types of residential energy use reduction program administrators. While there are advantages and disadvantages to each of these groups serving as program administrators, existing programs continue to reach only a limited number of individuals who have the potential to reduce energy in their home, leaving room for additional groups to help incentivize further residential energy use reduction.

Utilities

Utilities typically offer a wide variety of programs to encourage homeowners to implement energy efficiency in their homes, such as free lightbulbs, rebates for efficient appliances, and free or discounted energy audits (Eto et al., 2002). For example, Duke Energy Carolinas' Smart $aver program provides a combination of rebates, home assessments, and other benefits to help customers implement energy efficiency (Duke Energy, 2015). Advantages to utilities playing this role include their industry expertise and direct access to customers.

While utility-sponsored energy efficiency programs have continued to grow, regulators and interest groups have observed that utilities' financial incentives are not well aligned with the goal of reducing energy use (Holburn & Vander Bergh, 2006). Utilities typically earn profit from both the sale of energy and their capital investments, both of which are decreased by reductions in energy use. Therefore, many utilities face a disincentive to help customers reduce their energy use (Smith, 2015). Second, with respect to rooftop solar energy, many utilities are concerned not just with the lost energy sales but also with the cost of updating infrastructure to account for two-way delivery of electricity and the intermittency of solar energy (NRDC, 2012).

Nonprofit Entities

Recognizing that utilities face competing goals with respect to energy conservation, energy efficiency, and distributed energy, in some states nonprofit organizations play a critical role in administering energy use reduction programs that might otherwise be administered by a utility. Such groups as the EnergyTrust of Oregon (http://energytrust.org), for example, exist primarily to administer energy use reduction programs to help community members save energy and generate renewable energy. These groups' incentives and goals often align directly with emission and energy reductions within their communities. In addition, their ability to mobilize at the grassroots level can create significant momentum that can help overcome the barriers to implementation. However, there are challenges to creating an effective and sustainable nonprofit to administer these types of programs, including the need to create a new organization where no appropriate and trusted group already exists, building trust and relationships within the community, and maintaining a sustainable funding stream (Blumstein et al., 2005).

State and Local Agencies

Finally, state and local governments can and do play a role in administering energy use reduction programs, such as the federally funded and state administered Weatherization Assistance Programs that provide air sealing, insulation, lighting retrofits, and other services to eligible low-income community members (Maryland Department of Housing and Community Development, 2015).

While state and local agencies often have goals that align with the goals of energy use reduction programs, agencies can also be pulled in multiple directions, making it difficult to prioritize these efforts. Competing priorities and political pressure can mean that insufficient resources are allocated to energy use reduction programs. In addition, there are bureaucratic challenges to implementing these types of programs within state and local governments, including restrictions on hiring and inefficient procurement policies (Blumstein et al., 2005).

Criteria for an Ideal Energy Use Reduction Program Administrator

Blumstein and colleagues identified four characteristics that energy efficiency program administrators should have that can be applied to the energy use reduction pyramid as a whole: an internal alignment with the goal of energy use reduction, an incentive structure that supports success of the program, the ability to scale, and the ability to develop program infrastructure that will last (Blumstein et al., 2005).

In addition, other research points to three key ingredients to success in implementing the program at the ground level: the trust of individuals within the community, a direct line of communication to them, and resources to provide basic support (time and/or money). Trust is essential because residents are being asked to take action within their own homes and with their own money. Due to the personal aspect of these decisions, residents will hesitate to take action if the encouragement and information do not come from a trusted and apparently impartial source (Corner & Randall, 2011). In addition, a direct line of communication can help mobilize residents to take action and also help guide them through energy reduction programs. Without constant and clear communication, it can be easy for residents to lose momentum and not follow through with implementation, especially after the initial energy audit or solar assessment (Hall et al., 2013, pg. 4563-64). Finally, the ability of a program administrator to provide staff time to support the program and/or monetary resources is critical for the success of any program (Shehu & Akintoye, 2009).

Through our experience with our own pilot programs, discussions with administrators of similar programs, and general research, we propose that one particular group has many, if not all, of these characteristics and the opportunity to make a significant impact in this space but has yet to materialize as a significant player in the arena of residential energy use reduction programs: the employer. Employers are in a unique position to help employees overcome many of the barriers discussed in this chapter through the development and implementation of energy use reduction programs, and employer energy use reduction programs could be an excellent complement to the myriad programs currently in place across the country.

Home Energy Use Reduction as an Employee Benefit Program

Employers are uniquely positioned in that they possess many of the attributes of an ideal program administrator. First, many employers already have human resources and sustainability departments that are working to improve employee satisfaction and corporate sustainability, which may align well with the goals of an employee home energy use reduction benefit program. Second, employers often have the trust of their employees and can fill the trusted advisor role, vetting and recommending approved energy auditors and contractors as well as financial partners and other stakeholders. In this way, employers can help provide trusted information and education to help employers overcome the information barrier.

Third, employers typically have direct lines of communication to their employees, further facilitating the sharing of trusted information, including the ability to have informational sessions during work time and provide resources through online channels such as email and the company intranet. By providing educational information on energy efficiency and renewable energy directly to the employee as part of the work day, rather than requiring research during off-work time, employers are able to partially address the time and convenience barriers as well.

Finally, large employers may be particularly well situated to assist with home energy use reduction. For example, large employers span many regions and may have the opportunity to reach many employees. Larger employers may even be able to leverage group purchasing power to secure discounts for their employees. Employers with sufficient resources may have the ability to assign a department or position to help streamline processes for employees—answering questions that arise and providing a neutral expert opinion. Some employers may even be able to subsidize resident purchases directly, acquire group purchase discounts, and/or set up a low-interest loan program in partnership with local banks and credit unions.

Why Employers Should Consider Energy Use Reduction Programs

Beyond being uniquely positioned to facilitate energy use reduction measures among employees, many of the benefits of developing and implementing these types of programs fall closely in line with employers' needs. Specifically, there are two primary reasons that employers should consider creating and implementing employee home energy use reduction benefit programs. First, many employers already have voluntary commitments to reducing greenhouse gas emissions, and energy use reduction programs can help them meet these goals. Second, programs that help employees implement different tiers of the energy reduction pyramid can serve as strong employee benefits to help attract and retain employees.

Meeting Greenhouse Gas Emissions Targets

While the United States has made a political commitment to reduce greenhouse gas emissions nationally, many organizations have also developed emission reduction commitments at a more local level. One recent development is large employers' setting emission reduction goals as part of their corporate sustainability goals. In particular, two groups—corporations and universities—stand out as having many voluntary employer-based emissions reductions programs. Table 7-2 provides examples of such programs.

Table 7-2. Employer greenhouse gas reduction targets

Organization	Reduction Target(s)
Google	Became climate neutral in 2007; aims to use 100 percent renewable electricity
Microsoft	Became climate neutral in 2012; employs an internal carbon tax on each metric ton of carbon dioxide equivalence
Chevrolet	Reduce a total of 8 million metric tons of carbon dioxide equivalence emissions through the funding of carbon offset projects
Harvard University	Reduce emissions by 30 percent below 2006 levels by 2016; reduce per-capita water use by 50 percent below 2006 levels by 2020
Duke University	Become climate neutral by 2024
Cornell University	Become climate neutral by 2035
ABB	Reduce greenhouse gas emissions by 1 percent each year from fiscal year 1998 through fiscal year 2005
Alcoa	Reduce greenhouse gas emissions by 25 percent from 1990 levels by 2010, and by 50 percent from 1990 levels over the same period if Alcoa's inert anode technology succeeds
BP	Reduce greenhouse gas emissions by 10 percent from 1990 levels by 2010
Shell	Reduce greenhouse gas emissions by 10 percent from 1990 levels by 2002

Note: Reduction targets were found on each organization's website.

Existing employer greenhouse gas reduction commitments are often incorporated into a company's overall corporate social responsibility and public relations programs. To reach their commitments, employers typically reduce some emissions onsite, but many also purchase renewable energy credits or carbon offsets to increase their impact. An employee energy use reduction program represents another potential opportunity to reduce emissions offsite to count toward an existing greenhouse gas reduction goal.

In addition, many employers prefer to implement greenhouse gas emission reduction programs that have additional economic, social, and environmental co-benefits (benefits beyond the environmental benefit of greenhouse gas emissions reduction). Employers may place high value on local greenhouse gas emissions reduction programs that can directly benefit the local community and provide positive public relations. Employer programs that help employees reduce energy align well with the goal of achieving greenhouse gas reduction commitments through initiatives that have significant economic, social, and environmental co-benefits within the local community. These types of programs could also prepare employers for potential federal or state greenhouse gas mandates through early experience with the clean energy pyramid.

Providing a Unique Employee Benefit

In addition to reducing greenhouse gas emissions, these programs can also serve as unique employee benefits, helping residents to increase their standard of living by saving money, protect against risk of future energy price increases, increase comfort within their homes, and increase indoor air quality. While certain benefits are mandatory—such as Social Security unemployment insurance, worker's compensation, and family medical leave (Employee Benefits Research Institute [EBRI], 2011)—according to a national compensation study conducted by the US Bureau of Labor Statistics (2015), the most common benefits offered in the workplace today are voluntary benefits such as vacation pay, holiday pay, and health care. The Employee Benefits Research Institute (EBRI, 2011) identifies the common benefits shown in Table 7-3.

Table 7-3. Common workplace benefits

Mandatory
- Social Security retirement (OASI)
- Social Security disability (DI)
- Medicare Part A (Social Security HI)
- Workers' compensation
- Unemployment insurance
- Medicaid
- Supplemental Security Income (SSI)
- Public assistance

Voluntary—Tax Deferred
- Keogh plans
- Defined benefit pension plans
- Defined contribution retirement plans (pension, 40l(k), 403(b), stock ownership plans)

Other Tax Preferred
- Life insurance
- Long-term disability insurance
- Sick leave or sickness and accident insurance
- Other leave (maternity, funeral, jury, etc.)

Voluntary—Tax Exempt
- Employee and dependent health insurance*
- Retiree health insurance
- Dental insurance
- Vision insurance
- Medicare Part B (Social Security SMI)
- Educational assistance
- Child care
- Discounts
- Flexible spending accounts
- Parking
- Cafeteria facility
- Meals

Voluntary—Fully Taxable
- Vacations
- Paid lunch
- Rest periods
- Severance pay
- Cash bonuses and awards
- Legal assistance

Source: Employee Benefits Research Institute (EBRI, 2011).

Employers have numerous incentives to provide unique and valuable employee benefits. According to the Employee Benefits Research Institute, employee benefits help to promote the economic security of the employed by providing insurance against uncertain events and raising living standards for employees. This can be especially important during times when wages remain stagnant. ERBI also notes that benefits can help attract and retain employees, thereby reducing costs associated with turnover. In addition, employee benefits can provide financial incentives to reward employees' work, provide educational opportunities and professional development, increase employee loyalty, improve morale, maximize job performance and productivity, provide work-life balance, communicate the employer's mission and values, and build trust in the employer-employee relationship (EBRI, 2011).

As employers compete to retain top employees, they are becoming increasingly creative in the benefits and incentives they offer. Unique benefits such as stocked break rooms, on-location fitness centers, financial planning, childcare, and subsidized commuting can have a significant effect on an employee's finances and morale and help build employee loyalty. As Larry Page, CEO of Google, stated in an interview with *Fortune Magazine*:

> When you treat people [well], you get better productivity. Rather than really caring what hours you worked, you care about output. We should continue to innovate in our relationship with our employees and figure out the best things we can do for them. ... Our people have also been a lot happier and more productive, which is much more important. (Lashinsky, 2012)

Top employers, including many found on the Forbes List of Best Places to Work in 2015 (Dill, 2014), continue to develop and offer employees nontraditional benefits that are of high importance to the employees and their families. An employee benefit program centered on the energy reduction pyramid can provide many of the advantages that ideal benefit programs provide. Because the employer-employee relationship is so important to the success of an employee's work-life balance, employers are in a unique position to help employees reduce energy use and improve the health of their homes.

Case Study: The Duke Carbon Offsets Initiative (DCOI) Pilot Programs

Since its inception, the DCOI has been exploring emission reductions through the development of pilot programs in swine waste-to-energy, energy efficiency, renewable energy, and urban forestry. Currently, two of these pilot programs are exploring how residential energy efficiency and rooftop solar can provide social and economic benefits to the surrounding communities while also providing emissions reductions. These pilots test the ability of Duke University to leverage its status as a major employer in the region to remove the barriers to energy efficiency and renewable energy discussed earlier in this chapter. For more information on these programs, visit the Duke Carbon Offsets Initiative website, http://sustainability.duke.edu/carbon_offsets/.

As discussed previously in this chapter, energy conservation, energy efficiency, and distributed renewable electricity each have unique barriers that prevent people from taking respective action. In addition, each category is not readily available to all residents of a community. The DCOI program model ensures that each tier is available to each employee and that every employee has immediate access to at least one tier. For example, even with a group discount and tax credits, the average cost of a residential solar installation is approximately $8,000, which may be too expensive for some employees. In contrast, everyday changes in behavior can significantly reduce home energy use and can be done at a minimal cost and should be available to most employees.

The overarching goal of these pilots is to develop a scalable employee program that reaches all levels of Duke employees through the full energy use reduction pyramid. In addition, energy use reductions and renewable energy production from this program would result in greenhouse gas reductions that could be used by Duke University to help achieve climate neutrality.

In order to design an effective, scalable program that reaches the entire energy use reduction pyramid, it is important to implement smaller pilot programs that test each tier individually. As part of this process, the DCOI has designed and implemented two employee pilot programs: the DCOI-HEAL energy efficiency pilot that works with employees to increase their home's energy efficiency through retrofits and the Bass Connections solar group purchasing pilot that aims to make solar more accessible to employees through discounts. Below we share findings from these pilot programs and show how an employer can play an important role in reducing greenhouse gas emissions from electricity use at the residential level.

Energy Efficiency Retrofits—DCOI-HEAL

The DCOI-HEAL pilot program assists employees through the energy efficiency retrofit process by providing an educational presentation on energy efficiency retrofits, a no-cost Building Performance Institute (BPI) certified energy audit, assistance with energy audit scheduling, a personalized energy plan summary of the energy audit results, a list of vetted contractors, access to a low-interest loan rate from the Duke University Federal Credit Union, and a follow-up assessment to ensure quality retrofit work was completed. Each piece of the program is designed to remove the barriers to implementation outlined in this chapter. To address lack of information, the DCOI hosted information sessions for all participants. To increase trust in contractors, the program vetted and checked afterward on the quality of the work. In partnership with a local credit union, participants gained access to low interest loan products. To reduce the burden of time and inconvenience, program staff helped with scheduling appointments with contractors so participants did not need to take a lot of personal time. Finally, the DCOI is collecting energy data from all employees who participate to track the average energy use reductions from energy efficiency retrofits.

Residential Solar Pilot—DCOI Bass Connections in Energy

To complement the DCOI-HEAL pilot program, the DCOI developed a residential solar program for employees through a Bass Connections in Energy project. This project brought together students, staff, and faculty to explore the current solar landscape in North Carolina, develop educational documents for employees, and launch a Solarize Duke campaign to provide employees with a solar group discount and access to seasoned local solar installers with a track record in previous group Solarize projects. The DCOI residential solar program has aimed to remove as many barriers to implementation as possible. Similar to the DCOI-HEAL program, the DCOI hosted informational sessions for employees to learn about residential solar. The DCOI Bass Connections team also created a web resource to help employees understand the basics of residential solar, including two student reports and links to local solar resources. Both DCOI staff members and Solarize representatives were available to answer additional questions from participants via phone or email.

The DCOI also partnered with NextClimate, a local nonprofit that organizes and runs solarize programs, to access their local solar installers to improve participant trust in contractors. Staff reduced costs by allowing employees to pool with employees from other local companies in the Research Triangle region to receive a group purchasing discount based on the total amount of solar installed. While these savings are dependent on the size system a participant installs and the total amount of solar installed, typical savings range from $650 to $1700 (Pinder, 2015). Staff also saved participants time through a streamlined process for getting a solar assessment and proposal from participating installers. Lastly, solar installers have collected energy production data and provided it to the DCOI.

Lessons Learned from the DCOI Pilot Projects

The development and implementation of these pilot programs has provided the DCOI with firsthand experience of the challenges of executing energy use reduction programs. Following is a discussion of these challenges and potential ways to address them in the future.

Time Intensive

Both pilot programs proved to be time intensive for team members. The DCOI-HEAL program required significant amounts of time from DCOI staff to schedule audits, summarize audit results, and present the results to the employee. In particular, developing the personal energy plans required significant preparation in addition to the time needed to present the report. While less overall time was needed for the residential solar pilot, significant time was still required to organize, market, and present at events. The main issue associated with these significant time requirements was the fact that the DCOI staff members involved had other projects to manage in addition to these pilot programs. Thus, time management was challenging and priorities often had to be shifted.

One way to address this challenge would be to assign one staff member to manage these programs as his or her primary job responsibility. This would help prevent time management/prioritization issues and allow that person to specialize in and become more efficient at managing the program. Another option would be to use technology to automate certain parts of this process. Audit scheduling and summaries could both potentially be handled by a web application. Video presentations could be recorded and made available online. Finally, certain parts of the program could be cut out if staff time were not available to handle the entire program.

Tax Implications

Any benefits provided to employees are potentially subject to federal and state taxes. In the DCOI-HEAL program, paying for the audits was considered supplemental income for employees, increasing their tax burden. To account for this, the DCOI adjusted the benefit in order to pay for the taxes associated and provide a true no-cost audit for pilot participants. This almost doubled the cost of each audit to the DCOI. For the residential solar program, there were no tax implications because the DCOI acquired access to the Solarize program for employees at no cost to the University.

To date, the DCOI has not identified a straightforward solution to address this challenge. The DCOI has explored the potential for a third party to provide these benefits to Duke employees, but this framework could still require tax payments on any employee benefits provided. The main strategy to address this challenge is to limit the monetary value of these benefits in order to limit the cost to the employer. Another option is to require the employee to take on the tax burden, but this may significantly decrease the value of the benefit to the employee.

Cost to Implement

Building on the challenge of tax implications, the monetary cost to implement a program can vary greatly depending on the design of the program. The DCOI-HEAL program cost around $1500 per employee to implement (not taking into account program development costs). In comparison, the Residential Solar Pilot program was implemented at a very low cost (only the cost of the educational events).

For programs that subsidize employee energy use reduction efforts, one option for managing these costs is to cap the total number of employees able to participate annually.

Liability Concerns

There are many potential liability concerns that could affect the employer if not addressed properly. For example, a participant could have a poor experience with the program or with a contractor. Or, damage could occur to a participant's home during retrofits or solar installations. To address these potential issues, the DCOI first designed the program in a way to minimize these risks. This included carefully selecting contractors for the programs and developing contracts with these contractors that protected Duke University and the employees. However, there will always be risks associated with implementing these types of programs, and to address this, the DCOI worked with Duke University General Counsel to develop participation waivers for both programs that clearly addressed potential scenarios and released the University from liability.

Available Contractors

The DCOI was able to identify a significant number of high-quality contractors for the DCOI-HEAL program. Similarly, NextClimate was able to identify a number of high-quality solar contractors. However, in smaller communities it is common to have a limited number of highly qualified contractors available for these types of programs. In these scenarios, employers may have to work with third parties to train contractors in the area to provide the services needed to implement energy use reduction programs. While this can be resource intensive, it can also provide the employer with the opportunity to help spur economic development and job growth in the region. Another option is for the employer to partner with a national company that provides retrofit or solar installation services to bring them to the region.

Claiming Emissions Reductions

The DCOI collects all energy use reduction data associated with these pilot programs in order to calculate total emissions reductions. The DCOI hopes to use these emissions reductions to help Duke University reach its climate neutrality goal in 2024. However, state utility policy creates the potential for double counting—the scenario where both the local utility and Duke University count these emissions reductions separately toward their own goals. In this scenario, double counting could lead to fewer emissions reductions overall since the same emissions reductions are counted twice. The DCOI has yet to find a solution to this challenge and will use the energy data collected to continue working to develop one.

Every state has different utility policies in place that affect this scenario differently. For example, South Carolina does not have a renewable energy portfolio standard that would require the local utilities to count the emissions reductions from these types of programs, thereby reducing the potential risk of double counting. If federal requirements for emissions reductions are passed, this will further complicate the situation.

Nonetheless, many corporations and universities with voluntary emissions reductions commitments are in states where this is not an issue. If this challenge is identified early in the program development process, these groups may have the flexibility to design and implement these programs in ways that do not conflict or compete with local utilities. In addition, there is potential for these groups to work with local utilities to implement these programs together, with each taking a portion of the emission reductions and applying them to their own commitment or requirement.

Additional Tools for Implementing a Successful Program

In addition to the lessons learned from the DCOI pilot programs, there are a few tools that can be used by employers to develop effective energy use reduction programs. Specifically, employers can use creative financing, gamification, and data collection to remove barriers to implementation and provide employees with a robust energy use reduction benefit.

Tools for Success—Financing

Some employers may have the unique ability to help their employees acquire financing that might otherwise be unavailable to them. These mechanisms can reduce overall interest rates and/or reduce the credit score needed to take out the loan. In addition, they can help employees at lower salary levels gain access to energy efficiency retrofits, thereby decreasing their energy costs and increasing their standard of living. The following comparison of four options summarizes the types of financing mechanisms—identified by the DCOI (Weiss & Vujic, 2014)—that could be used by an employer to fund residential energy efficiency projects.

Self-financing using cash and grants. A commonly used financing mechanism for energy efficiency is the use of existing sources of cash.

Green revolving loan fund. The investments from the green revolving loan fund are paid back either through energy savings—or a portion of the savings—that result from each project or from the repayment of the loan by the party that receives funds from the revolving loan fund. These payments are returned, or "revolved," back into the fund and can be used to finance new projects.

Credit enhancements. A number of credit enhancement mechanisms—including interest rate buy-downs and loan loss reserve funds—have been designed to use public or private funds in a way that lowers the effective cost of capital for borrowers and gives financial institutions more reassurance that they will be repaid.

Crowd-funded alumni investment program. Crowd-funded investment programs rely on small amounts of capital investment from a large number of individuals.

Employers must be careful to balance the cost of these financing mechanisms with the goal at hand. For example, using an interest rate buy-down to drop the rate from 10 percent to 5 percent on a five-year $5,000 unsecured loan would cost $713 per loan. Thus, scaling the program could prove expensive if employees took advantage of this program. If an employer is looking to create a substantial employee benefit, it might be willing to invest a significant amount in such a program. Alternately, if an employer is looking to reduce greenhouse gas emissions locally at a reasonable cost, it might have to cap its investment in the program. For example, Duke University's energy efficiency pilot project, discussed earlier, will calculate the cost per unit of emission reduced by the project by tracking the total cost of the project and dividing it by the total emissions reductions from reduced electricity use.

Tools for Success—Gamification

One way to provide employees with the information needed and encourage personal motivation is through gamification—creating a program that is both fun and educational while aiming to promote real-world behavioral change. Research on gamification discusses how it can be used to meet goals of energy and resource conservation within companies, schools, and communities.

One study analyzed the programs and results of 53 games that were used to "influence behavior around energy efficiency and sustainability" (Grossberg et al., 2015). On average, the games produced a 3 to 6 percent reduction in energy use during the program period. However, if the program is not designed carefully, this energy reduction may not continue in the long run and can regress to the baseline after the game is complete (Orland et al., 2014). Well-designed programs have clearly defined goals, easily measurable metrics, and a rewards program that encourages further energy use reductions.

Typically, a game encourages participation through team building and competition. For example, at Duke University the Green Devil Smackdown is a program that uses gamification to encourage students, staff, and faculty to incorporate sustainable practices in their lives. In 2014, over 900 Duke University students, staff, and faculty participated in the eight-week challenge, where teams earned points for using alternative transportation, volunteering at the Duke Campus Farm, and meeting the requirements for the Green

Workspace Certification program. This program successfully brought together a diverse group of participants from a wide range of departments around Duke.

Due to the success of previous sustainability-focused programs, it is evident that a large group of engaged Duke University employees are committed to helping the university reduce emissions. Similarly, gamification could be used by other large employers across the nation. In particular, gamification could be used not only to mobilize employees to take energy reducing actions, but also to help move employees up the energy use reduction pyramid shown in Figure 7-1.

Tools for Success—Energy Data Tracking

In 2009, the federally funded stimulus plan funded many energy efficiency retrofit program throughout the United States. These programs helped promote energy efficiency within local communities (US DOE, 2012b). For example, North Carolina received $132 million in funding for weatherization assistance programs (US DOE, 2010b).

Speaking locally to groups that received funding and implemented weatherization programs (e.g., City of Durham, Town of Cary, Town of Chapel Hill, Town of Carrboro), one of the common challenges for each program was data collection. Without proper data collection, it is difficult to determine the real effects of energy efficiency work. In addition, collecting electricity bills from individuals by hand is time-intensive and makes data entry and management difficult. Without the ability to track and analyze results, programs are unable to assess their effectiveness.

If good data are collected, programs can analyze their results by detailing real energy savings adjusted for weather, identifying the types of retrofits that are most effective, and identifying which housing characteristics affect the outcome of weatherization work and solar installation. Together these methods also allow programs to share best practices and continue making improvements.

To address this issue, employers can purchase web-based data collection programs that automate data collection and analyze the data collected. By collecting monthly energy and weather data in addition to tracking the types and costs of retrofits completed by participants in the pilot program, the DCOI and ResiSpeak hope to provide real energy saving numbers at a level of detail not seen before.

Conclusion

With the need for reducing greenhouse gas emissions growing more urgent every day, we cannot ignore the potential for employer-based energy use reduction programs to have a significant impact on reducing greenhouse gas emissions. However, these programs must be designed and implemented carefully to provide the greatest benefit to both employers and employees (Table 7-4). These types of programs are not suitable for all employers, and it is important that employers weigh the benefits and challenges.

Table 7-4. Benefits and challenges of employer-based energy use reduction programs

Potential Benefits	Potential Challenges

• Environmental stewardship
  – Reduction in local greenhouse gas emissions
• Providing a unique employee benefit
  – Increased employee morale and trust
  – Reduced turnover
• Supporting the local community
• Positive public relations
	• Managing liability
	• Tax implications
	• Providing resources—time and money
	• Data collection
	• Locating skilled local contractors
	• Claiming emission

These potential benefits and challenges will be unique for each employer, determined by factors such as the employer's location, policy landscape, and employee population. As long as the employer is aware of the potential challenges in implementing an energy use reduction program, however, many of these obstacles can be removed through proper program design.

It is important to note that the DCOI pilot programs only scratch the surface of what we can learn regarding employer energy use reduction programs. As more programs are developed, it is crucial that data be collected from these programs and shared between them in order to continue refining best practices. In addition, best practices should be shared between different types of program administrators (utilities, nonprofits, state and local governments, and employers) to determine how these programs can best complement each other. Finally, as technology continues to advance, further research should be undertaken to determine how new forms of technology can be used within existing and future programs to help overcome the barriers to energy use reduction.

Chapter References

American Council for an Energy-Efficient Economy (ACEEE). (2013, March). *Overcoming market barriers and using market forces to advance energy efficiency*. Retrieved from http://aceee.org/files/pdf/summary/e136-summary.pdf

Blumstein, C., Goldman, C., & Barbose, G. (2005, May). Who should administer energy-efficiency programs? *Energy Policy, 33*(8). Retrieved from http://www.sciencedirect.com/science/article/pii/S0301421503003458

Borgeson, M., Zimring, M., & Goldman, C. (2012, August). The limits of financing for energy efficiency. *ACEEE Summer Study on Energy Efficiency in Buildings*. Retrieved from American Council for an Energy-Efficient Economy, http://aceee.org/files/proceedings/2012/data/papers/0193-000155.pdf

Clean Energy Durham. (2014). Pete Street: Neighbor-to-neighbor energy education and savings. Retrieved from http://www.petestreet.org/

Clinton Climate Initiative. (2015). Home Energy Affordability Loan (HEAL) program. Retrieved from https://www.clintonfoundation.org/clinton-presidential-center/about/heal

Corfee, K., Cullen, J., Graham, S., Fry, N. R., Davis, B., Meyer, A., , . . Bloch, C. (2014, May 28). *California Solar Initiative: Third-party ownership market impact study*. Retrieved from California Public Utilities Commission, http://www.cpuc.ca.gov/NR/rdonlyres/55A4BF20-875A-4B40-AD7C-3C768104211E/0/CSIThirdPartyOwnershipImpactReportFINAL.pdf

Corner, A., & Randall, A. (2011, August). Selling climate change? The limitations of social marketing as a strategy for climate change public engagement. *Global Environmental Change, 21*(3). Retrieved from http://www.sciencedirect.com/science/article/pii/S0959378011000793

Dill, K. (2014, December). The best places to work in 2015. *Forbes*. Retrieved from http://www.forbes.com/sites/kathryndill/2014/12/10/the-best-places-to-work-in-2015/

DSIRE. (2015). Database of state incentives for renewables & efficiency. Raleigh, NC: NC Clean Energy Technology Center. Retrieved from http://www.dsireusa.org/

Duke Energy. (2015). Smart $aver program. Retrieved from http://www.duke-energy.com/north-carolina/savings/smart-saver.asp

Employee Benefit Research Institute. (2011, March). *EBRI databook on employee benefits: Employee benefits in the United States.* Retrieved from http://www.ebri.org/publications/books/?fa=fundamentals

Environmental and Energy Study Institute. (2013, July). South Carolina co-ops release results of "Help My House" energy efficiency pilot. Retrieved from http://www.eesi.org/press-releases/view/south-carolina-co-ops-release-results-of-help-my-house-energy-efficiency-pi

Eto, J., Goldman, C., & Kito, M. S. (2002, February 7). Ratepayer-funded energy efficiency programs in a restructured electricity industry. *The Electricity Journal, 9*(7). Retrieved from http://www.sciencedirect.com/science/article/pii/S1040619096802909

Feldman, D., Barbose, G., Margolis, R., James, T., Weaver, S., Darghough, N., …Wiser, R. (2014, September 22). Photovoltaic system and pricing trends: Historical, recent, and near-term projections (2014 ed.) [PowerPoint Slides]. Retrieved from http://www.nrel.gov/docs/fy14osti/62558.pdf

Frederiks, E. R., Stenner, K., & Hobman, E. V. (2015, January). Household energy use: Applying behavioral economics to understand decision-making and behavior. *Renewable and Sustainable Energy Reviews, 41.* Retrieved from http://www.sciencedirect.com/science/article/pii/S1364032114007990

Freehling, J. (2011, August). Energy efficiency finance 101: Understanding the marketplace. American Council for an Energy-Efficient Economy. Retrieved from http://aceee.org/sites/default/files/pdf/white-paper/Energy percent20Efficiency percent20Finance percent20Overview.pdf

Gillingham, K., Newell, R. G., & Palmer, K. (2009). Energy efficiency economics and policy. *Annual Review of Resource Economics, 1*, 597-619.

Grossberg, F., Wolfson, M., Mazur-Strommen, S., Farley, K., & Nadel, S. (2015). Gamified energy efficiency programs. American Council for an Energy-Efficient Economy. Retrieved from http://aceee.org/sites/default/files/publications/researchreports/b1501.pdf

Hall, N., Romanach, L., Cook, S., & Meikle, S. (2013). Increasing energy-saving actions in low income households to achieve sustainability. *Sustainability, 5.* Retrieved from http://www.mdpi.com/2071-1050/5/11/4561

Haynes, G. A. (2009, February 9). Testing the boundaries of the choice overload phenomenon: The effect of number of options and time pressure on decision difficulty and satisfaction. *Psychology & Marketing, 26* (3). Retrieved from http://onlinelibrary.wiley.com/doi/10.1002/mar.20269/pdf

Holburn, G. L. F., & Vanden Bergh, R. G. (2006). Consumer capture of regulatory institutions: The creation of public utility consumer advocates in the United States. *Public Choice*, 126. Retrieved from http://link.springer.com/article/10.1007/s11127-006-4317-y#page-1

Intergovernmental Panel on Climate Change. (2014). Climate change 2014: Synthesis report summary for policymakers. Retrieved from http://www.ipcc.ch/pdf/assessment-report/ar5/syr/AR5_SYR_FINAL_SPM.pdf

Lashinsky, A. (2012, January 19). Larry Page: Google should be like a family. *Fortune*. Retrieved from http://fortune.com/2012/01/19/larry-page-google-should-be-like-a-family/

Maryland Department of Housing and Community Development. (2015). Weatherization Assistance Program (WAP). Retrieved from http://www.dhcd.state.md.us/website/Programs/wap/Default.aspx

Natural Resources Defense Council (NRDC). (2012, May). Removing disincentives to utility energy efficiency efforts. Retrieved from http://www.nrdc.org/energy/decoupling/files/decoupling-utility-energy.pdf

North Carolina General Assembly. (2009). House Bill 512. Retrieved from http://www.ncleg.net/Sessions/2009/Bills/House/PDF/H512v5.pdf

North Carolina Cooperative Extension at North Carolina State University. (2015). Home energy conservation. Retrieved from http://energy.ces.ncsu.edu/

Orland, B., Ram, N., Lang, D., Houser, K., Kling, N., & Coccia, M. (2014). Saving energy in an office environment: A serious game intervention. *Energy and Buildings, 74*. Retrieved from http://www.sciencedirect.com/science/article/pii/S0378778814000747

Pinder, R. (2015, April 8). What is Solarize Duke? Retrieved from http://www.solarizenc.org/howitworks?duke

Rai, V., & McAndrews, K. (2012, May). Decision-making and behavior change in residential adopters of solar PV. In C. Fellows (Ed.), *Proceedings of World Renewable Energy Forum (WREF) 2012*. Boulder, CO: American Solar Energy Society. Retrieved from https://ases.conference-services.net/resources/252/2859/pdf/SOLAR2012_0785_full percent20paper.pdf

Shehu, Z., & Akintoye, A. (2009). The critical success factors for effective programme management: A pragmatic approach. *The Built & Human Environment Review, 2*. Retrieved from http://www.tbher.org/index.php/tbher/article/view/9

Smith, R. (2015, April 20). Utilities' profit recipe: Spend more. *The Wall Street Journal*. Retrieved from http://www.wsj.com/articles/utilities-profit-recipe-spend-more-1429567463

Sorrell, S., Schleich, J., Scott, S., O'Malley, E., Trace, F., Boede, U., . . . Radgen, P. (2000, September). Reducing barriers to energy efficiency in public and private organisations. Science Policy Research Unit at University of Sussex. Retrieved from http://www.sussex.ac.uk/Units/spru/publications/reports/barriers/final.html

Sovacool, B. K. (2009, April). The importance of comprehensiveness in renewable electricity and energy-efficiency policy. *Energy Policy, 37*(4). Retrieved from http://www.sciencedirect.com/science/article/pii/S0301421508007556

United Nations Framework Convention on Climate Change (UNFCCC). (2015). About UNFCCC [webpage]. Retrieved from http://newsroom.unfccc.int/about/

US Bureau of Labor Statistics. (2015, March 11). Employer costs for employee compensation news release text. Retrieved from http://www.bls.gov/news.release/ecec.nr0.htm

US Census Bureau. (2015). National summary report and tables: 2013 [Data file]. Retrieved from http://www.census.gov/programs-surveys/ahs/data/2013/national-summary-report-and-tables---ahs-2013.html

US Department of Energy. (2010a). Homeowners guide to financing a grid-connected solar electric system. Retrieved from http://www1.eere.energy.gov/solar/pdfs/48969.pdf

US Department of Energy. (2010b). North Carolina Recovery Act state memo. Retrieved from http://energy.gov/sites/prod/files/edg/recovery/documents/Recovery_Act_Memo_North_Carolina.pdf

US Department of Energy. (2012a). LED lighting. Retrieved from http://energy.gov/energysaver/articles/led-lighting

US Department of Energy. (2012b). Success of the Recovery Act. Retrieved from http://www.energy.gov/sites/prod/files/RecoveryActSuccess_Jan2012final.pdf

US Environmental Protection Agency. (2015a). Draft inventory of US greenhouse gas emissions and sinks: 1990-2013. Retrieved from http://www.epa.gov/climatechange/pdfs/usinventoryreport/US-GHG-Inventory-2015-Main-Text.pdf

US Environmental Protection Agency. (2015b). Energy Star: Save energy at home. Retrieved from http://www.energystar.gov/index.cfm?c=products.pr_save_energy_at_home

Weiss, J., & Vujic, T. (2014). Financing energy efficiency-based carbon offset projects at Duke University. Retrieved from http://www.efc.sog.unc.edu/sites/www.efc.sog.unc.edu/files/Financing percent20Energy percent20Efficiency percent20Offsets.pdf

Zeelenberg, M., & Pieters, R. (2002). Beyond valence in customer dissatisfaction: A review and new findings on behavioral responses to regret and disappointment in failed services. *Journal of Business Research, 57* (4). Retrieved from http://www.sciencedirect.com/science/article/pii/S0148296302002783

CHAPTER 8

Increasing the Effectiveness of Residential Energy Efficiency Programs

Laura Langham, Amit Singh, Sarah Kirby, Sarah Parvanta, Marina Ziemian, and Donna Coleman

Introduction

A considerable amount of energy is wasted in the residential sector due to ineffectively built homes and inefficient occupant behavior (Seryak & Kissock, 2003; Stein & Meier, 1999; Zimring et al., 2012). Numerous intervention strategies have been implemented to address this issue, including home energy audits, energy use monitoring, weatherization, adoption of energy saving behaviors, and resident education. Each approach has its merits, but using only one of these strategies may limit the overall impact of home energy interventions. To achieve the goal of lowering consumers' energy use, a successful program will have greater impact if the dwelling and the resident are viewed as an energy-consuming system. Successful programs will also emphasize behavior-change interventions that are designed to reduce energy use (McLean-Conner, 2009).

Furthermore, home energy programs that use comprehensive approaches, integrated program delivery, and partnerships, rather than strategies with a singular focus, present more opportunities to change energy use behavior (McLean-Conner, 2009). Thus, while offering a one-dimensional solution such as home audits has its merits, intervention effectiveness increases when combined with other strategies (Abrahamse et al., 2005). Reducing residential energy consumption and improving efficiency in existing homes requires interventions in both occupant behavior and building efficiency. In addition, research demonstrates that "participating customers realize large benefits above and beyond the basic energy savings they enjoy from programs" (Skumatz et al., 2000, p. 8.363).

This chapter provides a brief overview of currently existing residential energy efficiency programs and takes an in-depth look at three specific programs (case studies) that implement various strategies toward achieving reductions in residential energy efficiency. The chapter concludes with recommendations to energy efficiency program administrators on best practices for future energy efficiency program modeling.

Understanding the Landscape of Existing Energy Efficiency Programs

Energy efficiency programs vary greatly in scope, methodology, focus, and outcomes. Here we review different facets of energy efficiency programs: program sponsors, types of programs, program services, funding mechanisms, and strategies for targeting program participants.

Program Sponsors: Regional and local utilities, municipalities, community organizations, educational institutions, nonprofits, government agencies, and businesses offer energy efficiency programs. The services offered and funding structures vary widely, depending on the goals and origins of the programs. Often, two or more entities with aligned goals will partner, developing a program that increases program capability and meets the desired outcomes of both organizations. For example, local governments often partner with utilities and community organizations to support weatherization of homes for low-income residents. Utility companies and municipalities often partner with Energy Star to increase impacts of energy savings or to leverage funds. Collaborative efforts between a university cooperative extension program and state energy office have worked to extend financial and human resources to provide greater reach for educational efforts.

Types of Programs: A report by the Consortium for Energy Efficiency (CEE, 2014) provides a good overview of existing residential energy efficiency programs. The report focuses on 50 programs implemented by 48 CEE members, mostly electric utilities. The individual programs are separated into two broad categories: comprehensive home performance approaches, and prescriptive measure upgrades. Of the two approaches, the comprehensive home performance approach is the most involved. It is a whole-system approach that looks at the existing energy systems within a home and how those systems interact with each other. This information is then analyzed to create a detailed work plan that achieves the greatest energy savings for the home over a long period of time. The comprehensive approach relies on certified contractors who utilize best practices from nationally recognized standards. One example of a program that utilizes the comprehensive approach is Home Performance with Energy Star.

The prescriptive measure upgrades approach is less involved than the comprehensive home performance approach. This approach comprises a series of individual program components or system upgrades and does not necessarily include a complete assessment of a home's energy use.

Of the 50 programs highlighted in the CEE report, 20 used only a prescriptive approach, 12 used only a comprehensive approach, and 18 used both a prescriptive and comprehensive approach. The programs were established starting in 1978, but 23 of the programs (or nearly half) were established after 2008. Besides homeowners, the programs also target contractors, realtors, property owners, installers, developers, and auditors. Thirty-six states, the District of Columbia, and one Canadian province are represented in the report, with a collective budget of over $500 million.

Public benefit programs: Some states require investor-owned utilities to collect a surcharge from ratepayers that is used to fund programs such as energy efficiency, renewable energy, or low-income energy assistance. While utilities typically administer public benefit programs, in some states the fund is administered by a nonprofit or contractor organization. Examples of this latter approach include the New Jersey Clean Energy Program, Wisconsin Focus on Energy, and the Energy Trust of Oregon. Local governments also may have an opportunity to access these funds directly to implement local energy efficiency initiatives (US Environmental Protection Agency, n.d.).

Cooperative extension programs: Examples of programs that serve all residents in a state or region are cooperative extension energy education programs. Cooperative extension programs are housed at land grant universities across the nation; their mission is to transform research from the university into practical application for citizens. These programs are managed at a state or regional level and typically focus on citizen education through workshops, online resources, and outreach efforts. Some of these programs have extended their impact by securing grants or contracts with government organizations in order to offer additional services such as home energy audits and weatherization services (see the Case Study: E-Conservation Residential Energy Program later in this chapter).

Types of Services Provided to Program Participants: Regardless of the program, each provides services to program participants. Depending on the specific program, services often include one or more of the following: community education, energy audits, rebates, low-interest loans and incentives, marketing and sales of energy efficiency products and services, appliance recycling, direct install, and retrofits. Successful energy efficiency programs incorporate a portfolio that encompasses more than one of these services (Abrahmse et al., 2005).

In order to properly identify a home's energy efficiency needs, a home energy audit or assessment must be completed. An energy audit is typically defined as a comprehensive home evaluation, complete with diagnostic equipment such as a blower door test for identifying air leakage, infrared camera to identify and record heat loss, and/or duct blaster, designed to identify duct leakage. A home assessment, sometimes called a "clipboard audit," is less rigorous than an audit and is a walk-through visual inspection of a home. Most programs, though not all, require auditors to be Building Performance Institute (BPI) and/or Home Energy Rating System (HERS) certified (LeBaron & Renaldi, 2010). Home assessments and audits both typically provide a homeowner report listing recommended upgrade measures. In more advanced reports, information such as potential energy savings, expected improvements to comfort and/or health, and the required initial investment is also provided (McEwan, 2012).

Retrofits are a service for existing homes that implement a suite of energy efficiency improvements in one concentrated effort, with appropriate measures determined using building science techniques to optimize homes' performance (CEE, 2010). As the centerpiece of many residential energy efficiency programs, retrofits are a proven strategy for improving home performance, thereby reducing residential energy consumption and improving indoor air quality. They provide environmental, economic, health, and other social benefits, potentially reducing a resident's energy usage 20 to 40 percent. Growing in numbers, programs offering retrofits "exist in every region of the country, and have flourished in a range of climatic zones, in very different social and legislative environments, and with a range of sponsors, predominant types of energy, and energy costs" (LeBaron & Renaldi, 2010, p. 2). Important steps to save residential energy include sealing ducts, insulating basements/attics, adding weatherstripping, upgrading heating equipment, and adding programmable thermostats (Granade et al., 2009).

Funding: Programs are most often funded through utility rate payment, which builds program costs into customer utility rates or utilizes government grants, municipal initiatives, or grassroots fundraising. In addition, a small but growing number of small businesses have developed programs, usually incorporating the sale of products or services to maintain viability.

Targeting Program Participants: Program participants are identified and determined based on program sponsors' targeted population and specific goals. While some programs cast a wide net, offering services to all residents in a certain region or state, most are available to specific utility customers, a community, or, in other cases, are based on income and need. A utility company, for example, may offer an incentive program to all customers or may design a program that focuses on a subpopulation of its customers. When defining and targeting program participants, it is important for the program designers to clearly define the goals of funders and carefully consider which participant group best meets those goals. The targeted group of participants will vary, depending on program impact goals. Is the goal simply to lower utility use among participants, or is there a social justice component to the program? For example, according to the National Research Council,

> residential energy use varies by household income… Upper-income households earning more than $100,000 annually in 2001 used about twice the energy used by lower-income households earning under $15,000 annually. But the energy burden (the fraction of income spent on energy) is much higher for lower-income households compared with middle or upper-income households. (2010, p. 47)

While sponsors, funding, scope, targeted residents, and services vary widely among programs, all are contributing to our understanding of best practices in reaching the common goal of increasing residential energy efficiency. The US EPA, for example, maintains a database of programs and information that on the one hand are intended to assist state and local governments in developing policies and programs to improve energy efficiency, and on the other hand to target customers of all types directly with information that can be used to make sound decisions about their energy use (US EPA, n.d.). Resources include examples of building codes for energy efficiency, customer incentive programs, guidance on evaluating energy efficiency programs, and best practices for providing energy use and cost information.

Case Studies

Case Study One: Austin Energy Residential Power Saver Program

The Austin Energy Residential Power Saver Program is an example of a program that provides a home analysis to participants who wish to take advantage of Home Performance with Energy Star rebates and financing options. As a ratepayer program, Austin Energy's Power Saver Program finances more than $17 million in energy efficiency incentives and rebates annually by building the program costs into electricity rates (Home Performance Resource Center, 2010).

The success of this program is due, in part, to a city ordinance, the Energy Conservation and Disclosure ordinance, which requires an energy audit of all homes over 10 years old before they can be placed on the market. By requiring this ordinance, the city of Austin raises awareness about energy efficiency and its importance in determining a home's value (Institute for Market Transformation, 2011). Prospective buyers utilize the audit reports as an informational tool to make a more informed decision about homes they are considering purchasing. Because the audit requirement falls to the home seller, energy efficiency upgrades that are identified in the report are less likely to be addressed because the owner/seller will not be benefiting from the energy savings.

In order to qualify for the financial incentives, homeowners must select a program-approved contractor to conduct an efficiency analysis that provides the owner with retrofit recommendations along with a cost estimate for the recommended changes. The analysis is a visual inspection that lasts about 30 minutes and does not incorporate diagnostics such as blower door or infrared testing (Home Performance Resource Center, 2010). Once the home analysis is complete, the homeowner seeks bids to complete the recommended work and submits the selected bid to Austin Energy. At this stage, the homeowner may choose one of two incentive options through the program: (1) take out a loan, or (2) receive rebates (Climate Leadership Academy Network, 2010).

The rebate option tends to be appropriate for making small upgrades to the home, while the loan option can help cover the costs of more extensive upgrades. If opting for a loan, home retrofitting work begins after Austin Energy approves the contractor bid and, in turn, secures the loan. The loans are low-interest, unsecured energy improvement loans that do not require a lien on the property. Rebates, covering the lower of $1,575 or up to 20 percent on the improvement cost, are mailed to the homeowner within four weeks of approval of designated retrofits (Peterson et al., 2011).

Program Strengths

The success of this program is due to factors such as its funding structure, strong partnerships, the offering of rebates and financing options, and the implementation of the City of Austin Energy Conservation and Disclosure ordinance. By building program costs into the electricity rates of all customers, Austin Energy has a steady and predictable stream of funding. Partnerships with Home Performance with Energy Star, Austin Institute of Real Estate, Austin Community College, and area contractors have built and strengthened program capacity. The program has also resulted in significant workforce development as local contractors have had to increase their staffing to meet the growing demand. Austin Community College provides training programs for area energy auditors and, for those who are actively serving Austin Energy customers, there is financial support for approved courses.

Another strength of this program is the offering of low-interest loans or rebates to participating homeowners. For homeowners who wish to make smaller retrofits, the rebate program lowers the financial hurdle. For owners who wish to make more substantial energy efficiency improvements to their home, there are two loan options to choose from, allowing qualified participants to get an unsecured loan that is locked in at interest rates ranging from 0 percent APR for 3 years to 6 percent for 10 years. By offering these rebate and loan options to participants, Austin Energy is providing a variety of options for homeowners to select a path that best fits their energy efficiency needs and financial situation.

The City Energy Conservation and Disclosure ordinance increased the effectiveness of the program by requiring all sales of homes older than 10 years to receive a home efficiency analysis and to provide the report to prospective homebuyers. The requirement of this report raises awareness about residential energy efficiency as a factor in choosing a home.

Opportunities

While this program is accomplishing great success, additional program components may add to the impacts that can be realized. By providing educational opportunities to citizens, the program can help increase awareness and knowledge. Education can also help bridge the gap between knowing and acting by helping consumers better understand the rewards of increased energy efficiency. Workshops that educate all citizens about how occupant behavior impacts energy usage can increase the savings of all citizens, regardless of their ability to engage in a loan program. For those homeowners who participate in the loan or rebate program, impacts will likely increase if they are educated about how their home works as a system, what they can do themselves without hiring a contractor, and the variety of no-cost things they can do to lower their energy use. In addition, continued engagement with participants through workshops, newsletters, additional incentives, and community forums can keep former participants engaged, furthering their efforts to reduce the amount of energy they consume.

Case Study Two: Consumer Education Program for Residential Energy Efficiency

One approach to motivating residents to reduce their energy consumption is through education. An example of a program that incorporates residential energy efficiency education to encourage energy conservation is the Consumer Education Program for Residential Energy Efficiency (CEPREE). This program was created in 2003 through a partnership between the Cornell Cooperative Extension (CCE) and the New York State Energy Research and Development Authority (NYSERDA). CEPREE leverages the education expertise of Cornell and NYSERDA with the extensive network of the CCE to form an educational collaborative that can access much of New York's population (Laquatra et al., 2009). The goal of this program is to transform the energy efficiency market by creating a demand among consumers and housing providers.

Cornell Cooperative Extension, through its statewide network of educators, has the ability to reach citizens in every county of the state of New York. NYSERDA is a state agency whose goal is to "promote energy efficiency and the use of renewable energy sources" (NYSERDA, 2015, para. 1). NYSERDA accomplishes this goal largely through partnerships with stakeholders, such as the one it has with the CCE.

Since the program's inception in 2003, CEPREE has reached over 70,000 New Yorkers in 44 counties (Laquatra et al., 2009). The program's target audience is homeowners, renters, and builders and other housing professionals. CEPREE reaches its audiences in a variety of ways including presentations at county fairs, workshops, and mass media campaigns. For example, the program has a portable Energy Bike that it can take to county fairs or environmental events at schools (Laquatra et al.). The bike enables users to power electrical devices by pedaling, allowing them to see how much energy is required to run different appliances. Laquatra et al. also outline CEPREE's Energy Town Meetings, which connect the general public to energy experts from Cornell or NYSERDA. These town meetings are generally held at county offices and last for about an hour, followed by question and answer sessions.

In addition to education of the general public, CEPREE conducts events for more specialized audiences. In collaboration with a number of organizations, Cornell faculty delivered an education speaker series in 2004 to 133 homebuilders in five different locations around New York (Laquatra et al., 2009). The speaker series was focused on resources and information related to high-performance homes and Energy Star homes. Laquatra and colleagues state that CEPREE has also educated owners of apartment buildings about NYSERDA resources available to help them conduct energy efficiency upgrades on their properties. CEPREE educators have even reached out to retailers to encourage them to stock more Energy Star appliances (Laquatra et al.).

While there has not been sufficient funding to gather detailed and specific information about the behavioral changes that have resulted from CEPREE's programs, a preliminary report from 2006 noted that people who attended presentations on residential energy efficiency reduced their annual electricity bills by $400 and their annual carbon dioxide emissions by an average of 2.52 metric tons (Laquatra et al., 2009). Additionally, Laquatra et al. (2009) report that a CCE 2006 system-wide survey showed that 69 percent of people who attended CEPREE's residential energy efficiency workshops went on to participate in the recommended Home Performance with Energy Star program. This adoption rate could translate to a number of other energy and greenhouse gas reduction benefits attributable to the CEPREE program.

Program Strengths

The CEPREE program has developed strong partnerships, effectively targeted audiences, and achieved steady funding. CEPREE has been able to connect New York's citizens with energy experts from Cornell and NYSERDA, providing a trusted and unbiased source of information for residents. The Cooperative Extension Service (i.e., CEE) has a strong track record of successful program implementation. The partnerships between these organizations have led to innovative and impactful educational programs. In tailoring its programs to various stakeholders, CEPREE has been able to give specific recommendations to different populations. CEPREE has worked with homebuilders, apartment building owners, and retailers, among others, to reduce residential energy consumption. Finally, by partnering with NYSERDA, CEPREE has a steady income source, a Systems Benefit Charge. This monthly charge on the electricity bills of New York ratepayers is a large funding source for NYSERDA's partnerships and programs.

Opportunities

Preliminary results show that CEPREE's initiatives have been successful in encouraging New York residents to reduce their home energy consumption. As the program continues, it will need to search for opportunities to continue to engage with homes to further efficiency upgrades. While education is certainly an important part of getting residents to reduce their energy consumption, research has shown that combining education with other strategies can increase effectiveness.

Another opportunity for CEPREE is to fund a detailed and comprehensive program evaluation. By undertaking a thorough evaluation, CEPREE would be able to identify the most effective strategies in order to better direct its resources.

Case Study Three: E-Conservation Residential Energy Program

In early 2000, the US was undergoing a renewed interest in residential energy efficiency, yet in North Carolina, consumer energy education and awareness programs were lacking. It was in this environment that the E-Conservation Residential Energy Education program was created to bridge the gap between housing research, advancement in energy efficiency technologies, and consumer education and implementation. The program's mission was to reach residents and provide them with the knowledge, tools, and experiences that would lead to the adoption of energy efficiency behaviors and measures, thus reducing energy consumption (Kirby et al., 2015; Kirby et al., 2014). This case study examines the development of the E-Conservation Program between 2004 and 2015 as it adapted program strategies in response to consumer needs.

Energy efficient appliances and technologies have increased considerably over the years. Efforts such as Home Performance with Energy Star, Building America, the Better Buildings Neighborhood Program, and the Partnership for Advancing Technology in Housing have all focused efforts on creating efficient, high-performance housing. However, there is a disjuncture between housing research and the translation of the research into consumer adoption of technology in homes.

In order for market transformation to occur, specifically in the retrofit market, consumers need "consistent, accessible, and trusted information" (Middle Class Task Force and Council on Environmental Quality, 2009, p. 5) in order to make informed energy decisions. The purpose of the E-Conservation Residential Energy Education program is to address these concerns by increasing consumers' knowledge about ways to save energy and to teach them "to be proactive in reducing home energy consumption and in saving money through low/no cost energy efficiency measures, energy technologies, behavioral changes, and home energy retrofits" (Kirby et al., 2009, para. 3).

Implementation Strategies

The E-Conservation program implemented several strategies to educate citizens about energy efficiency and provide them with tools that would lead to a decrease in residential energy use. In the beginning, strategies focused on educational workshops, energy kit distribution, and residential audits. Homeowners attended an educational workshop that provided strategies about ways to increase efficiency through no-cost and low-cost measures. These face-to-face workshops provided the opportunity to engage residents in meaningful conversations where they were encouraged to share their experiences, identify personal roadblocks to implementing change, and receive individual strategies for overcoming those obstacles.

Workshops were structured to incorporate delivery of information, hands-on learning, and group interaction. These workshops were followed up with a brief survey, asking the participants to identify impacts such as behavior change, implementation of strategies, and installation of energy kit items such as CFLs and high-efficiency showerheads (Kirby et al., 2008; Kirby et al., 2015). In more recent iterations of the program, additional and continued engagement with workshop participants was achieved through monthly newsletters that provided next steps, how-to instructions for home efficiency projects, and promoted utility and government rebates and incentives.

In addition, the program has included a limited number of home energy audits. To qualify for an audit, individuals must first attend an energy education workshop. Participants who received a home audit were provided with report about their home's energy efficiency needs, along with recommended measures that could be taken to increase the home's efficiency. While this approach did lead to behavioral changes and decreased energy usage, evaluation of the program determined that a large number of the participants did not take action in making their homes more efficient. The primary reasons residents gave for not acting on the audit recommendations were (1) lack of money, (2) lack of time, and (3) lack of information (Kirby et al., 2008; Kirby et al., 2014). While participants acknowledged that the recommendations would increase energy savings, they encountered barriers to taking action on the recommendations.

As a result, in subsequent years, the program incorporated home retrofit measures into the strategies to help residents achieve their energy efficiency goals, conducting 250 home assessments and retrofits between August 2014 and May 2015. The requirement remained for participants to attend an educational workshop that engaged attendees through multimedia presentations and hands-on activities demonstrating how to effectively implement energy-saving strategies and measures in existing residences. The resident was also required to be present for the home audit and retrofit, shadowing the contractor. In this way, the homeowner learned about his or her home's specific energy needs and how to improve the home's efficiency. Additionally, while the contractor was in the home, the homeowner completed $600 in energy upgrades to the home. This new strategy ensured that the program had an impact on the energy efficiency of each home because specific actions were taken at the time of the audit. Even if the homeowner chose not to or was unable to make additional changes based on audit recommendations, the home was made more efficient by the direct services provided at the time of the audit.

Another feature that was added to the program was an upgrade of the audit tool used for the program. Previously, data from the audit reports were entered into reports manually by program staff. This cumbersome, labor-intensive system was improved greatly by using a mobile reporting app for audits and home reports. The new app allows contractors to easily input data, take photos, draw diagrams, and complete the homeowner report at the time of the audit. The report is directly downloaded into a database that is accessible to program staff, eliminating effort, increasing accuracy, and enhancing the reports.

Measuring Savings by Collecting Utility Data

The E-Conservation program took on the task of verifying program participants' energy savings by collecting actual energy usage. In order to calculate actual energy savings, the program needed to collect two years of past utility usage data and two years of data post-audit. Obtaining utility usage data from participants or utility companies (or both) was time-consuming and often ineffective, leading to a loss of data. Ultimately, much of the utility data for the serviced homes were not collected due to lack of participant compliance. It was too difficult for participants to continue to provide utility usage information to the program for two years after the audit. This 18 to 24 months of data, however, provided vital information that was needed in order to accurately identify changes in usage. Even if the data were successfully obtained, entering the large amounts of data into a database required significant time and effort and required program support.

To improve this process, E-Conservation partnered with ResiSpeak, a tool that provides residents with their gas and electricity usage monthly, complete with weather normalization, graphs, and comparisons to past monthly and yearly energy usage. Once participants signed up with ResiSpeak through the E-Conservation online portal, both the resident and the program could access the home electric and gas usage. By incorporating the ResiSpeak utility access tool, the program management had access to monitor participants' energy usage and—just as important—participants could easily access and view their electric, gas, and water usage, all in one location. This increased access for participants provided continued engagement and helped heighten their awareness about their energy use. This component of the program increases its impact because "providing feedback on energy consumptions by including easy-to-understand comparative information on energy use on monthly utility bills," can positively influence consumer behavior (National Research Council, 2010, p. 291).

Disseminating and Distributing Information

There are many challenges in providing residents with the information they need to be equipped to make decisions and take action regarding energy use. In order to be successful in providing information that leads to behavior change, the program needed to provide accurate, credible, clear, unbiased, accessible, and practical information.

Because program participants are voluntary learners, it is critical that they have the desire, interest, and motivation to actively engage in the learning process. They need to understand why the information is important to them and how gained knowledge can positively impact them. Educational material was designed to be accessible to the typical consumer, taking large amounts of technical information and breaking it down into workshops, hands-on experiences, fact sheets, how-to efficiency projects, articles, and pamphlets that the average resident could relate to and understand.

When developing educational materials, the following guidelines were followed:

- Assume that participants have varying levels of confidence and skill in completing home improvement projects.

- Assume that participants have unequal skill in reading, writing, and math.

- Assume that Internet access is limited or unavailable for some.

- Ensure the information is unbiased and is based on the most current research findings.

- Create a positive learning environment where all levels of knowledge are acceptable and where questions are expected and valued.

- Clearly convey the relevance of the material to residents.

To increase the reach and impact of the program website, many new features were developed such as resources for renters, videos, how-to guides, information about available rebates and incentives, and an extensive dictionary of energy efficiency terms, complete with images and helpful links. Staff intended such online resources to educate residents about home energy, how the house works as a system, strategies for reducing energy use through changes in occupant behavior, how to reduce energy use through retrofits and upgrades, how to take advantage of available rebates and incentives, how to prioritize projects, and how to hire a contractor. Staff also have shared website material through social media tools such as Facebook, Twitter, and Pinterest. This strategy of developing new, meaningful resources has resulted in 13,000 unique views per month.

Program Strengths

The E-Conservation Program incorporates education, outreach, home assessments, retrofits, and utility usage access and feedback to engage participants in a variety of ways, addressing both occupant behavior and building efficiency. By combining strategies, the program is able to offer a variety of interventions geared to serve citizens in becoming more energy efficient. In addition, by partnering with utilities, city governments, organized citizen groups, and energy efficiency professionals, the program maximizes its capabilities.

Program workshops and online resources provide valuable information to residents about behavioral and retrofit strategies for reducing energy use and increasing energy efficiency in their home. While workshops and online resources are essential tools in motivating and informing citizens to live more sustainably, the program achieves the greatest impacts through combining a variety of educational strategies with residential energy assessments and retrofits. The program has created greater opportunities for contact with consumers through home energy workshops and more interactions with home energy raters through the audit process.

The program also disseminated consumer energy kits as an incentive to encourage participants to make simple energy changes. As described earlier, staff also distributed online resources, videos, workshops, residential audits and retrofits, and utility use educational tools to educate and empower residents to become more aware of their energy use. These tools assist consumers in developing strategies for lowering their energy demands, and implementing strategic energy efficiency changes that increase the efficiency of their homes.

Opportunities

Because E-Conservation is funded by the North Carolina State Energy Office, this program is heavily reliant on one funding source. Developing other sources of funding would increase the longevity and impact of this program. Further development of key partnerships would also enhance the existing program and could lead to additional development of services offered.

For example, at this time, the E-Conservation Program does not incorporate low-interest loan programs or rebates to assist homeowners in completing major retrofit measures to their homes. Though some retrofitting is completed for each participating home, most still have weatherization needs that extend beyond the program capabilities. Putting in place a loan or rebate component to the program would facilitate homeowners who desire additional retrofits for their homes, based on the home assessment offered by E-Conservation.

Lastly, in order to further assist residents, E-Conservation is in the process of developing avenues to continue participant support through an "Energy Efficiency Coach" model. Researchers have found that homeowners who receive personal individual communication from energy advisors or professionals give a higher priority to energy improvements than those who received only written educational leaflets (Nair et al., 2010). This continued support and guidance, the program administrators believe, will further increase impacts in their quest to assist residents to lower their utility use.

Strategies for Increasing Program Effectiveness

In this chapter we have presented information about several energy efficiency program initiatives and strategies and detailed a variety of effective and prevalent strategies. We also reviewed three programs that have used varied strategies to reduce energy use and increase energy knowledge: one focused on resident educational workshops and resources, one incorporating audits and incentives, and one that provides educational workshops and resources, energy kits, home assessments, retrofits, and access to utility usage data.

The energy audit is an opportunity for programs to educate residents about how their home works as a system, providing them with valuable information specific to their home and engaging them in the process. All too often, contractors do not involve the homeowner in the audit process. Approaching the energy audit as an educational opportunity for the homeowner can increase the impact of the audit.

Many homeowners do not understand how a house works as a system and do not have a clear understanding of their home's energy use deficiencies. A home audit report provides homeowners with a blueprint of how to make changes in their home in order to reduce their energy use, make the home more efficient, and increase overall comfort. Homeowners can obtain even greater knowledge and a clearer understanding of what can be achieved if the homeowner shadows the auditor, learning about the specific needs of their home. Following the auditor and allowing him or her to explain where energy loss is occurring is extremely beneficial.

For example, if a home audit report lists "lower your water temperature to 120 degrees," the homeowner may not act on that recommendation because he or she may not know how to complete this action. If, on the other hand, the contractor points to the water heater temperature guide and shows the homeowner the location of the settings, explaining the why and how, then a barrier to taking action has been removed. On a similar plane, if homeowners are told to clean their clothes dryer lint trap to avoid risk of fire and to help the machine run more efficiently, the impact will be increased if the homeowner actually sees the lint-packed vent.

Many homeowners understand that air leakage causes energy loss, but the how and where of air leaks is often a mystery to a homeowner. Having a contractor point out areas of air leakage or penetration and explain or demonstrate how to seal those leaks can demystify this energy loss, thus increasing the chances that the homeowners will take action to improve their home's efficiency.

While program goals vary from one residential energy efficiency program to another, all strive to reduce homeowner energy usage. Some programs have one interaction with a program participant, while others engage participants through several steps and interventions. All, however, at some point discontinue engagement with program participants. While there is little research on this topic, some research seems to indicate that continued engagement increases participant knowledge, commitment, behavior, and adoption of energy efficiency practices (McEwan, 2012).

Not all programs are able to monitor true energy use over time. In fact, according to the National Research Council,

> much of the available data on energy use in buildings is based on self-reporting or inferences rather than on direct measurement, and estimates of uncertainties around the data are seldom available. Expanded data gathering, particularly through direct measurement, would facilitate more rigorous evaluation of energy efficiency measures and would contribute to the accuracy and completeness of future studies. (2010, p. 44)

The ability for homeowners to monitor real time energy use provides them with information that may positively reinforce energy-saving behaviors and enable them to see the real-time impact of energy-saving technologies.

For some programs, once the energy audit is complete, representatives of the intervention program are no longer in direct and consistent contact with homeowners. Some homeowners may then reduce their engagement with efficient energy-use behaviors and lifestyle changes. A number of strategies are available to aid in maintaining homeowner engagement after the audit. Table 8-1 suggests ways to connect with homeowners after the main energy audit is complete in an attempt to continue their engagement and efforts toward achieving a more energy-efficient home.

Table 8-1. Dissemination channels and intervention strategies to increase homeowner engagement after an energy audit

Channels for disseminating energy program information to increase engagement	Intervention strategies to increase engagement

• Resources available on website
• Social media
• Newsletters
• Fliers
• Personalized coaching
• Workshops
	• Mobile technology notifications
	• Social media reminders
	• Text messaging
	• Mobile apps
	• Real-time energy use monitoring tools
	• Email reminders
	• Online or direct personalized coaching
	• Online workshops
	• Online home self-assessment tools
	   – Guides
	   – Videos
	• Financial incentives
	• Energy kit distribution

Note: These are suggestions only and are not based on our own empirical research.

Conclusion

After reviewing the variety of options available for continued engagement regarding energy use behaviors, we believe that the most effective programs will incorporate a variety of strategies from each program. Perhaps most relevant to homeowner engagement and retention are the variety of methods educators can use to directly communicate with homeowners after the main energy audit is complete, some of which are suggested in Table 8-1. Reviewing the home audit with the homeowner is another strategy for facilitating homeowners in making their homes more energy-efficient.

To facilitate the process even more, programs can incorporate coaching. Helping homeowners by reviewing the auditor's findings, creating a list of tasks to complete, setting priorities, making a timeline, and identifying rebates and incentives—in addition to providing instructional videos and other tools—can keep homeowners engaged, increasing the likelihood that they will implement the suggested retrofit measures.

The content of these interventions delivered through various channels can be presented in various forms. For example, mobile apps could offer reminders about air filter replacements and other home energy adjustments that need to be addressed periodically. Other online sources could provide workshop videos, personalized coaching, or a number of energy-tracking tools for use in the home. Finally, interventions could also offer financial incentives directly to homeowners for using energy-efficient products in the home, as well as energy kits for installing those products.

Although these ideas reflect strategies to increase homeowner energy engagement following the initial energy audit, they are not guaranteed to be effective. Without a formal evaluation or other empirical testing to determine the effects of these strategies on a population's engagement behavior, it is difficult to be certain of their utility. It is also critical to understand which communication channels a particular population of homeowners finds most accessible. Some homeowners may pay more attention to energy program fliers that arrive through the postal service versus energy program messages in their email or Twitter feeds, for example. Some homeowners may not have adequate interest in or access to the Internet, a smartphone, or other social media technology to make those channels viable. An assessment of channel preferences and feasibility should take place before an intervention is implemented to ensure that the intervention can in fact reach its intended audience.

Chapter References

Abrahmse, W., Steg, L., Vlek, C. & Rothengatter, T. (2005). A review of intervention studies aimed at household energy conservation. *Journal of Environmental Psychology, 25,* 27–291. http://dx.doi.org/10.1016/j.jenvp.2005.08.002

Climate Leadership Academy Network. (2010). *Case study: Austin, Texas using energy information disclosure to promote retrofitting climate.* Retrieved from http://web.mit.edu/cron/Backup/project/urban-sustainability/Energy%20 Efficiency_Brendan%20McEwen/Cities/Austin/austin_energy_disclosure.pdf.

Energy Star. (n.d.). Home performance with Energy Star [homepage on the Internet]. Retrieved from https://www.energystar.gov/index.cfm?fuseaction=hpwes_profiles.showsplash

Granade, H. C., Creyts, J., Derkach, A., Parese, P., Nyquist, S., & Ostrowski, K. (2009). *Unlocking energy efficiency in the US economy: Executive summary.* New York: McKinsey & Company, McKinsey Global Energy and Materials. Retrieved from http://www.mckinsey.com/client_service/electric_power_ and_natural_gas/latest_thinking/unlocking_energy_efficiency_in_the_us_ economy

Home Performance Resource Center. (2010). *Best practices for energy retrofit program design case study: Austin energy residential power saver program.* Retrieved from http://www.hprcenter.org/sites/default/files/ec_pro/ hprcenter/best_practices_case_study_austin.pdf

Institute for Market Transformation. (2011). *Third-party performance testing a case study of residential energy code enforcement in Austin, Texas.* Retrieved from http://www.imt.org/resources/detail/case-study-1-third-party-performance-testing

Kirby, S., Chilcote, A., and Guin, A. (2008). Residential energy audits: A tool for enhancing consumer energy efficiency and conservation improvements (pp. 75–78). In B. L. Yust (Ed.), *Proceedings of the 2008 Annual Conference of the Housing Education and Research Association*, Indianapolis, IN.

Kirby, S., Chilcote, A., and Guin, A. (2009). Energy education: Ideas that work. *Journal of Extension, 47*(5). http://www.joe.org/joe/2009october/iw2.php

Kirby, S.D., Guin, A.H. & Langham, L. (2015). Energy education incentives: Evaluating the impact of consumer energy kits. *Journal of Extension, 53*(1). Retrieved from http://www.joe.org/joe/2015february/rb5.php

Kirby, S.D., Guin, A.H., Langham, L., & Chilcote, A. (2014). Exploring the impact of the E-Conservation Residential Energy Audit Program. *Housing and Society, 41*(1), 53–70.

Laquatra, J., Pierce, M. R., & Helmholdt, N. (2009). The Consumer Education Program for Residential Energy Efficiency. *Journal of Extension, 47*(6). Retrieved from http://www.joe.org/joe/2009december/pdf/JOE_v47_6a6. pdf

LeBaron, R. & Renaldi, K.S. (2010). *Residential energy efficiency retrofit programs in the US: Financing, audits, and other program characteristics.* Washington, DC: National Home Performance Council.

McEwan, B. (2012). *Community based outreach strategies in residential energy upgrade programs* (master's thesis). Retrieved from http://web.mit.edu/ energy-efficiency/docs/theses/mcewen_thesis.pdf

McLean-Conner, P. (2009). *Energy efficiency—principles and practices.* Tulsa, OK: PennWell Corporation.

Middle Class Task Force and Council on Environmental Quality. (2009). *Recovery through retrofit.* Retrieved from https://www.whitehouse.gov/ assets/documents/Recovery_Through_Retrofit_Final_Report.pdf

Nair, G., Gustavsson, L., & Mahapatra, K. (2010). Owners perception on the adoption of building envelope energy efficiency measures in Swedish detached houses. *Applied Energy, 87.* 2411-2419.

National Research Council. (2010). *Real prospects for energy efficiency in the United States.* Washington, DC: National Academies Press.

New York State Energy Research and Development Authority (NYSERDA). (2015). About NYSERDA. Retrieved from http://www.nyserda.ny.gov/ About

Peterson, D., Matthews, E. & Weingarden, M. (2011). *Local energy plans in practice: Case studies of Austin and Denver* (Technical Report NREL/ TP-7A20-50498). Golden, CO: National Renewable Energy Laboratory, US Department of Energy. Retrieved from http://www.nrel.gov/docs/ fy11osti/50498.pdf

Consortium for Energy Efficiency (CEE). (2014). Overview of residential existing homes programs in the United States and Canada. Consortium for Energy Efficiency (CEE), Boston, MA. Retrieved from http://library.cee1. org/content/cee-2015-existing-homes-program-overview/

Seryak, J. & Kissock, K. (2003). Occupancy and behavioral affects on residential energy use. *Proceedings of the Solar Conference* (pp. 717–722). American Solar Energy Society; American Institute of Architects.

Skumatz, L. A., Dickerson, C. A., & Coates, B. (2000). Non-energy benefits in the residential and non-residential sectors—innovative measurements and results for participant benefits. In *Proceedings from the 2000 ACEEE Summer Study on Energy Efficiency in Buildings*. Washington, DC: American Council for an Energy-Efficient Economy. Retrieved from http://aceee.org/files/proceedings/2000/data/index.htm

Stein, J. R., & Meier, A. (1999). Accuracy of home energy rating systems. *Energy, 25,* 339–354.

US Environmental Protection Agency (EPA). (n.d.). Utilities and other energy efficiency program sponsors. Retrieved September 15, 2015, from http://www.epa.gov/statelocalclimate/local/topics/municipal-utilities.html

Zimring, M., Borgeson, M. G., Hoffman, I, Goldman, C., Stuart, E., Todd, A., & Billingsey, M. (2012). Delivering energy efficiency to middle income single family households (Report LBNL-5255E). Berkeley, CA: Lawrence Berkeley National Laboratory. Retrieved from http://middleincome.lbl.gov/reports/report-lbnl-5244e.pdf

Contributors

Following are the contributors and their affiliations at the time of writing.

Charles Adair, Duke University

Justin Baker, RTI International

Gibea Besana, RTI International

Donna Coleman , North Carolina Housing Finance Agency

Dan Curry, Clean Energy Durham

Elizabeth Doran, Duke University

Jason Elliott, Duke University

Nicholas Garafola, Duke University

Christopher S. Galik, Duke University

Melanie Girard, Grapevine Realty Services

Daniel Kauffman, Terracel Energy Consulting and ResiSpeak

Jim Kirby, Greenthinc

Sarah Kirby, North Carolina State University

Laura Langham, North Carolina State University

Ryan Miller, North Carolina Building Performance Association

Sarah Parvanta, RTI International

Laura Richman, Duke University

Douglas Rupert, RTI International

Amit Singh, Duke University

Claudia Squire, RTI International

Brian Southwell, RTI International, Duke University, and University of North Carolina at Chapel Hill

Kendall Starkman, Duke University

Jordan Thomas, Duke University

Joseph Threadcraft, Wake County Government

Jennifer Weiss, University of North Carolina at Chapel Hill

Gabrielle Wong-Parodi, Carnegie Mellon University

Marina Ziemian, RTI International

Index

Note: Page numbers followed by *t* and *f* indicate tables and figures. Numbers followed by *n* indicate notes.

39813195R00123

Made in the USA
Middletown, DE
26 January 2017